Basics of Radiation Protection for Everyday Use
How to achieve ALARA: Working Tips and Guidelines

Editors
Harald Ostensen
Gudrun Ingolfsdottir

Author
Leonie Munro

Artist Line Diagrammes
Merle Conway

Creative Digital Imaging
Fiona Walters

WORLD HEALTH ORGANIZATION

WHO Library Cataloguing-in-Publication Data

Munro, Leonie.
    Basics of radiation protection for everyday use : how to achieve ALARA : working tips and guidelines /
Editors: Harald Ostensen, Gudrun Ingolfsdottir ;
author: Leonie Munro.

1.Radiography - standards 2.Radiation protection 3.Radiation dosage 4.Quality control 5.Manuals
I.Ostensen, Harald. II.Ingolfsdottir, Gudrun. III.Title.

ISBN 92 4 159178 1   (NLM classification: WN 665)

© World Health Organization 2004

All rights reserved. Publications of the World Health Organization can be obtained from Marketing and Dissemination, World Health Organization, 20 Avenue Appia, 1211 Geneva 27, Switzerland (tel: +41 22 791 2476; fax: +41 22 791 4857; email: bookorders@who.int). Requests for permission to reproduce or translate WHO publications – whether for sale or for noncommercial distribution – should be addressed to Publications, at the above address (fax: +41 22 791 4806; email: permissions@who.int).

The designations employed and the presentation of the material in this publication do not imply the expression of any opinion whatsoever on the part of the World Health Organization concerning the legal status of any country, territory, city or area or of its authorities, or concerning the delimitation of its frontiers or boundaries. Dotted lines on maps represent approximate border lines for which there may not yet be full agreement.

> The mention of specific companies or of certain manufacturers' products does not imply that they are endorsed or recommended by the World Health Organization in preference to others of a similar nature that are not mentioned. Errors and omissions excepted, the names of proprietary products are distinguished by initial capital letters.

The World Health Organization does not warrant that the information contained in this publication is complete and correct and shall not be liable for any damages incurred as a result of its use.

The named authors alone are responsible for the views expressed in this publication.

Printed in Malta

List of editors, authors and collaborators

Editors
Harald Ostensen, M.D., Coordinator, Team of Diagnostic Imaging and Laborat
 Technology, WHO, Geneva, Switzerland, co-chairman, The Global Steerin
 Group for Education and Training in Diagnostic Imaging

Gudrun Ingolfsdottir, coordination and support, Team of Diagnostic Imaging a
 Laboratory Technology, WHO, Geneva,

Author
Leonie Munro, Assistant Director, Radiography, King Edward VIII Hospital,
 Durban, South Africa

Artist Line Diagrammes
Merle Conway, Durban, South Africa

Creative Digital Imaging
Fiona Walters, Medical Media Services: Nelson R Mandela School of Medicin
University of Natal, Durban, South Africa

Foreword

Modern diagnostic imaging offers a vast spectrum of modalities and techniques, which enables us to study the function and morphology of the human body in details that approaches science fiction.

However, it should also be remembered that thousands of hospitals and institutions worldwide do not have the possibilities to perform the most fundamental imaging procedures, for lack of equipment, malfunction or break down of equipment, or insufficient diagnostic imaging skills.

Therefore, WHO in collaboration with The International Commission for Radiological Education (ICRE) of the International Society of Radiology (ISR) is creating a series of manuals and workbooks developed under the umbrella of the Global Steering Group for Education and Training in Diagnostic Imaging. The main issue is to assist and guide "end-users" responsible for diagnostic imaging, be it radiologists, physicians, radiographers, nurses or others to improve safety and quality of their work.

The full series of manuals and workbooks will primarily cover basic examination techniques and interpretation of radiography and ultrasonography as well as radiation safety aspects and basic quality assurance issues .

The manuals are authored by authorities in the specific fields dealt within each manual, supported by a group of collaborators, that together cover the experience, knowledge and needs, which are specific for different regions of the world.

It is our sincere hope that the manuals and workbooks will prove helpful in the daily routine, facilitating the diagnostic work up and hence the treatment, to the best benefit for the patient

Geneva, Switzerland and Lund, Sweden, May 2004
Harald Ostensen
Holger Pettersson

Table of contents

Chapter 1 1
  Introduction 1
Chapter 2 3
  Production of X-rays 3
    Construction of anode for heat dissipation 5
    Selection of focus spot 6
    Basic X-ray unit components 7
    Risks and benefit for the use of ionizing radiation 8
    Biological effects of ionizing radiation 8
    Factors to minimize radiation dose to patients and staff 8
    Summary 10
Chapter 3 11
  X-ray rooms: design, materials and protection barriers 11
    Warning signs 11
    Occupancy of X-ray rooms: calculation of safety aspects 11
  Room size 12
    Protective cubicle 13
    Windows and air conditioning units 14
    Doors and walls 14
  Tips to ensure doors are closed during radiographic examinations 15
    Walls: materials and lead equivalence 15
  Proportions for barium plaster mix 17
    Ceiling and floors 17
    Ward radiography 17
    Change cubicles 18
    Safety measures in special procedure rooms 18
    Summary 18
Chapter 4 19
  Radiation protection devices 19
    Lead rubber aprons 19
    Protective lead rubber gloves 20
    Lead rubber gloves 21
    Thyroid shields 21
    Gonad shields 21
    Summary 22

Chapter 5 23
Beam-restricting devices 23
Use of lead blockers to improve image quality 25
Quality assurance tests of beam-restricting devices 25
Working tips 25
Collimator-beam alignment test 26
Method for checking the collimator-beam alignment 26
Test to check alignment of the centre of the X-ray beam 27
Method 27
Use of compression to reduce thickness of patient 29
Summary 29

Chapter 6 31
Scattered radiation: role of grids 31
Grid design 31
Grid ratio 32
Parallel and focused grids 33
Grid cut-off 34
Focused grid: importance of tube-side 36
Stationary grids 37
Moving grids 37
Scatter clean-up: role of grids 38
Care and maintenance of grids 40
Grid factors 44
Summary 44

Chapter 7 45
Radiographic technique, exposure factors, and quality assurance tests 45
Patient positioning 47
Selection of kV 50
Selection of mAs 51
Exposure manipulation: kV/mAs 61
Determining variable kVp charts 62
Quality assurance tests to minimize unnecessary film fog 67
Safelight tests 67
Processor control: performance monitoring 68
Careful film-handling and film storage 70
Summary 71

- **Chapter 8** 73
  - Exposure to ionizing radiation during pregnancy 73
    - Patients and medical staff 73
    - Pregnant radiation workers 73
    - Summary 73
- **Chapter 9** 75
  - Self-evaluation of images: application of ALARA 75
  - Practical hints for self-evaluation of image quality 75
    - Suggested answers to questions for Figures 9a to 9h 83

CHAPTER 1

Introduction

Ionizing radiation sources can cause harm both to humans and to the environment. The most important source of ionizing radiation is that used in medicine for diagnostic and therapeutic purposes. Several international guidelines and regulations have been published addressing this aspect of ionizing radiation. The most important of these publications is the *"International Basic Safety Standards for Protection Against Ionizing Radiation and for the Safety of Radiation Sources"*, Safety Series No 11 published by the International Atomic Energy Agency (IAEA), the World Health Organization (WHO), the International Labour Organisation (ILO) and other international organizations. A second publication of major importance addressing the same issues, is the *"1990 Recommendations of the International Commission on Radiological Protection, Publication 60"* published by the International Commission on Radiological Protection (ICRP). It is strongly recommended that these publications are made available to decision makers and relevant medical and technical staff.

It is important not to forget that the primary aim in radiography and radiology is to produce diagnostic images, which assist to establish a correct diagnosis, and thus be of benefit to the treatment of patients. Therefore, the image quality needs to be sufficiently good for diagnostic considerations, i.e., for pattern recognition.

In theory, optimal image quality allows one to make accurate diagnosis. Taking radiation dose into account, however, and to keep this in line with the ALARA principle (**A**s **L**ow **A**s **R**easonably **A**chievable), a certain "balance" between what is optimal and what is acceptable, may be needed. To repeat or not to repeat "sub-optimal" images depends on the clinical situation and the indications for performing an examination. When evaluating images, a decision to re-expose a patient is usually based on experience and a set procedure. If an image is unacceptable, then the radiation received by the patient was not justified, and certainly of no benefit.

The aim of this book is to provide guidance and tips to improve image quality without subjecting patients to unnecessary ionizing radiation

CHAPTER 2

Production of X-rays

In this chapter, the basics of the production of X-rays and risks and benefits of using ionizing radiation for examination of patients, are covered while focusing on the importance of radiation protection:

- A simple definition and explanation of the production of X-rays

- Examples of risks and benefits in terms of radiation dose to patients

- Characteristics and production of X-rays

Like visible light X-rays are part of the electromagnetic spectrum, but the wavelength is approximately 10.000 times shorter (Fig. 2a).

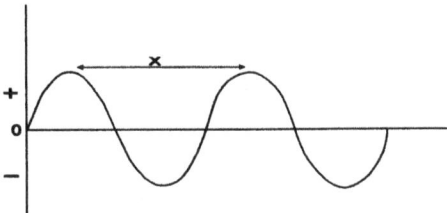

Figure 2a: Drawing of wavelength showing one wave=crest to crest (x).

The short wavelengths of X-rays give them specific properties:

- Ability to penetrate materials such as the human body.

- Ability to induce emission of visible light when hitting certain substances (fluorescence), a phenomenon responsible for reduction of radiation dose needed when using intensifying screens mounted in a light-tight cassette.

- When directed towards a photographic film, the silver halides in the film emulsion are converted to densities that become visible through film development ("processing").

- Ability to induce biologic changes in body cells and tissues, which is the principle behind radiotherapy.

- X-rays are produced in a so-called X-ray tube where the main parts are the cathode (negatively charged) with a filament, and an anode (positively charged).

By means of electrical current measured in milliampere (mA), the filament, which is comprised of coils of wire similar to a light bulb, is heated to glow just as we see in an ordinary light bulb. The difference with the X-ray tube filament is that it does not produce visible light but acts as a source emitting electrons when heated. As the temperature of the filament is raised (this is regulated by the mA settings), more electrons are emitted and the electrical current, or flow, through the X-ray tube increases. The duration of applying this current is expressed in seconds, and in radiography, the product of this current as regulated and measured in *ampere (or, milliampere)* and the time as measured in *seconds,* is called *exposure factor* and abbreviated mAs (milli-ampere-seconds).

The function of the *positive* anode is to attract the *negatively* charged electrons, which are produced by the filament of the cathode. The higher the electrical potential between the cathode and the anode is, the stronger this "attraction" of electrons (i.e., current) will be. The magnitude of this electrical potential, i.e., *difference,* between anode and cathode, is regulated by adjusting the *(kilo) voltage (kV)*.

During application of high voltage across the tube, the electrons impact (collide) with the angled anode and the following occurs:

- Great amounts of heat are produced.

- X-rays of varying wavelengths are produced when the rapidly travelling electrons decelerate (slow down) as they impact with the anode [Figure 2b].

- Ionizing radiation is produced.

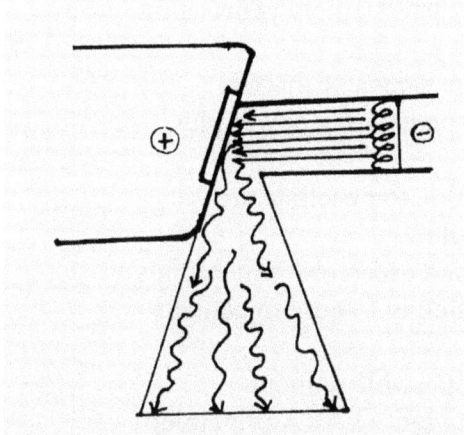

Figure 2b: X-ray tube with electrons flowing from the cathode to the anode. The X-rays are deflected from anode to pass through the tube window.

Construction of anode for heat dissipation

To be able to function properly for long periods (a long "tube life"), the anode must be able to withstand heat. The fast travelling electrons produce heat when they loose energy by impacting with the anode. Dissipation of the heat produced can be achieved by the use of suitable material in a tube.

- The material used in the construction of the anode is usually a block of copper to dissipate the heat.

- Additional material used is a tungsten plate set into the face of the anode in the centre of the tube. Tungsten is used as it has a high melting point allowing the anode to withstand very high temperatures when electrons strike it and X-rays are produced.

> **Tips to reduce dose: Selection of kVp and mAs**
> - The higher the kVp selected the more penetrating the beam = ↓dose.
> - Higher voltage results in X-rays of shorter wavelength and greater penetrating power plus greater intensity.
>
> ♦ Tip: Use highest kV possible to penetrate area of interest (an ALARA principle).
>
> - mAs (tube current x time in seconds/milliseconds) has direct role in contributing to dose to patients. Thus ↑mAs = ↑ dose.
>
> ♦ Tip: Keep mAs as low as possible without compromising image quality (an ALARA principle).

Selection of focus spot

X-ray tubes allow selection of different focus sizes, and the *focal spot* is the area of the anode which is bombarded by electrons from the heated filament. Built into the cathode is a *"focusing cup"*. Its function is to direct the electrons to an area of the tungsten target.

The size of the focal spot (source) has an important effect on the image formed as it is the slightly angled area of the target [Figure 2c] that is struck by the focused 'stream' of electrons. Thus, the smaller the area on the target is, the sharper the image will be.

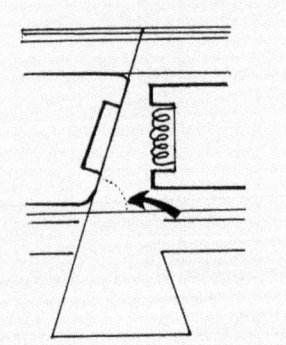

Figure 2c: Angle of anode (arrow).

The use of a small focus depends on the capabilities of the X-ray unit. A large focal spot can withstand more heat than a small one, but some image detail is lost. Focus size for each tube is determined by the manufacturer.

Basic X-ray unit components

A tube requires
- transformer that 'steps-up' the incoming voltage as most exposures include settings from 40k up to at least 150 kVp.

- rectifiers may be needed in countries where the electrical power is supplied as a single-phase alternating current.

- power supplies and controls for the filament, and timers to control duration of exposure.

- protective devices to reduce dose and also to prevent overheating of the tube.

Radiation safety in this context includes shielding of the tube to absorb X-rays emitted in all directions from the focus of the anode. Only the rays that pass through the patient are required for image formation. An X-ray beam consists of different energies of which most contribute to image formation. Low energy radiation, also called 'soft X-rays', do not contribute to image formation but add to the patient dose. Lead lining is used to absorb the majority of X-rays not contributing to image formation. 'Soft X-rays' must be reduced before the beam enters a patient. These low energy rays are absorbed by *inherent* and *additional* filtration. Inherent filtration is the modification of the X-ray beam from the anode. The beam is filtered as it passes through the tungsten deposit on the glass of the tube window, the oil used for cooling of the tube, and the shield aperture. This is specially covered in Chapter 5. Inherent beam filtration in a tube is a prerequisite for radiation protection.

Additional filtration is provided by a shield consisting of aluminium placed outside the tube aperture. The aluminium filter's role is to absorb most of the 'soft X-rays. Total filtration is the sum of the *inherent* filtration and *additional* filtration.

These safety components are mentioned very briefly because their proper functioning is an absolute prerequisite for adequate radiation safety. Regular basic tests should be performed as part of a quality assurance programme to ensure their proper functioning.

Risks and benefit for the use of ionizing radiation

The most important factor to consider is that no radiographic/radiological investigation should be done unless medically justified, i.e., there are sound clinical reasons for the examination, and that the benefit for the patient would outweigh possible radiation risks.

Biological effects of ionizing radiation

A simple explanation of these effects is that when X-rays pass through a patient they cause some biological changes or even damage. Potential damage to the body is of two types, namely *stochastic* and *non-stochastic,* also called deterministic.

- *Stochastic* means something that occurs as a result of the law of chance or probability, and is independent of radiation dose. Stochastic effects, due to exposure to ionizing radiation, can cause cancer, or have influence on gene-material affecting future generations.

- *Non-stochastic (= deterministic)* means that something will always occur, but only when exposure is exceeding a certain *threshold.* The degree of damage (severity) increases the more the threshold value is exceeded.

Factors to minimize radiation dose to patients and staff

Some radiation protection measurements such as filtration of the beam, rectification, and tube shielding are not within a radiographer's control. Others, and these are the main topics dealt with in this book, are within the control of the radiographer/operator, as shown below.

- limitation of field size to area of interest

- use of fast screen-film combinations whenever appropriate

- optimal film processing

- use of automatic exposure timers if available

- use of gonad shields
- selection of grid
- compression of obese patient
- highest practicable kV and lowest mAs
- reduction of number of repeats by careful patient positioning, and use of immobilization devices
- performance of basic quality assurance tests
- no continuous radiation during fluoroscopy
- only required staff allowed into room during radiographic examinations
- all staff should stand behind protective barrier during the exposure
- X-ray units must have adequate shielding
- staff who are required outside the barrier must wear lead-rubber aprons
- field size to be smaller than screen size during fluoroscopy
- staff should stand outside the path of the primary beam, and as far away from it as possible
- lead-rubber flaps to be used on image intensifiers to reduce scatter to staff

Safety measures to reduce dose to patients and staff should also be implemented in operating theatres, and angiography suites. Operators of fluoroscopy units/C-arms, etc., who are not trained in radiation protection measures, should *forced* by national laws to undergo basic training in radiation protection to avoid unnecessary dose to patients, staff, and the environment.

Summary

The production of X-rays for use in medicine could lead to slight chances of damage to living tissue. Provided protection measures are implemented, the risks of potential radiation-induced damage is minimal.

CHAPTER 3

X-ray rooms: design, materials and protection barriers

Most radiographic work is carried out in dedicated rooms (X-ray rooms). Radiation protection measures in these rooms are important because the people in the room, or in the close proximity, could be subject to ionizing radiation, either from the primary beam, or scattered off the patient, or the X-ray table. When considering room design and, all methods should be used to minimize unnecessary exposure to both primary and secondary (i.e. scattered) radiation. All X-ray rooms should allow access only to persons needed for the procedures. Those who work in the room should be protected by means of 'barriers'. These barriers should be thick enough to ensure that radiation dose received does not exceed internationally accepted limits for registered radiation workers.

Warning signs

Warning signs understood by both literate and illiterate persons, must be displayed. These signs must be placed at all entrances to X-ray rooms. In addition, warning lights automatically lit when ionizing radiation is produced, should be installed.

Occupancy of X-ray rooms: calculation of safety aspects

Room size, design, floors, ceiling, doors, and height of windows should be in accordance with national laws and generally accepted international radiation protection recommendations. Based on workload and occupancy factors, radiation safety aspects of X-ray rooms can be calculated by medical physicists [Figure 3a].

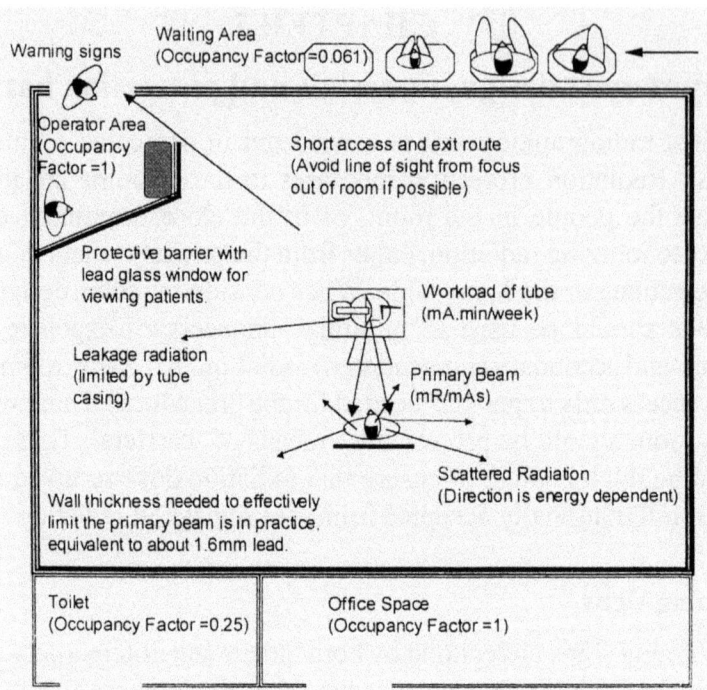

Figure 3a: Example of layout of an X-ray room showing safety aspects that must be considered when designing an X-ray room. Workload and occupancy factors are used by medical physicists to calculate the required thickness of the protective barriers. Protective barriers ensure workers in the room are protected from ionizing radiation. People (arrow) seated in the corridor in close proximity to the room must also be protected (W. Rae acknowledged for the diagram.)

Room size

It is important to note that the size of an X-ray examination room influences on radiation protection. The further one is from the primary beam (X-ray tube), the less radiation dose is received. This is strictly mathematical and based on the inverse square law [Figure 3b], i.e. an increase in distance by two metres reduces the dose by a factor of its "square" (in this case by 2 x 2 = 4).

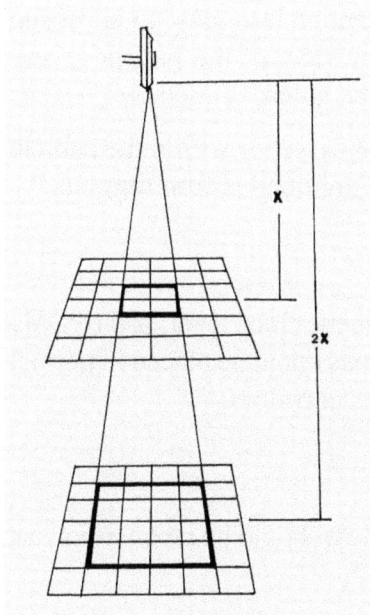

Figure 3b: Diagram showing principle of the inverse square law. As the distance from the source increases, the intensity of the beam decreases proportionally. Beam covers four squares at the distance X but sixteen squares when the distance from source is doubled (2X).

A general purpose X-ray examination room should not be smaller than 16 m^2 to allow safe and adequate installation of equipment. Often, the actual size of existing rooms cannot be altered. When so, proper implementation all radiation protection measurements becomes even more important.

Protective cubicle

The protective cubicle should allow sufficient space to accommodate staff while exposing the films [Figures 3c and d].

Figures 3c and 3d: Top diagram (3c) does not meet international standards of protection as the radiographer is not protected from primary and scattered ionizing radiation. Bottom diagram (3d) shows worker behind a fixed protective cubicle. Lead glass windows allow clear view of the patient during exposure.

The cubicle should be positioned in such a way that any radiation, be it direct or scattered reaching the radiographer/technologist operating the X-ray unit, is reduced to an absolute minimum. The cubicle should have at least one window with protective lead glass to allow the radiographer / technologist to have a clear view of the patient at any time, and the height of the cubicle should be at least 2 metres.

Storage of exposed and unexposed film cassettes within the cubicle during exposure is recommended to avoid unintended film fogging.

Windows and air conditioning units

These should be at least 2 metres above the floor level. If the X-ray room is above ground level, then the windows could be placed in normal height provided there is no link to passages/corridors.

Doors and walls

The position of doors is important as there should be no obvious risk of exposure to passers-by [Figures 3e and f].

- Sliding access doors give better radiation protection than normal doors; they should overlap each side of the door access/opening by at least 100 mm.

- Doors should be lined with lead sheet of 2 mm thickness.

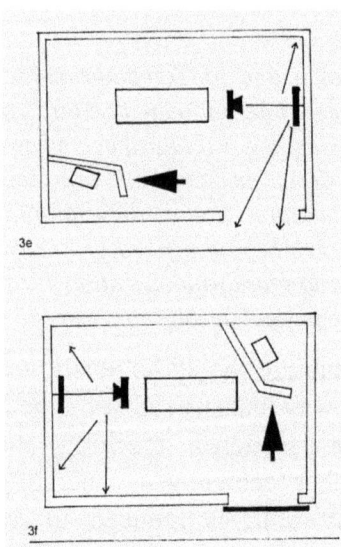

Figures 3e and 3f: Top diagram (3e) is an example of poorly designed room because the door is direct line of X-ray beam when erect Bucky is used. Bottom diagram (3f) is an acceptable design. The door is at opposite end of room out of way of X-ray beam. Protective cubicles in both diagrams have angled walls (thick arrows) to ensure radiographer / technologist is protected at all times from both primary and scattered radiation.

> **Tips to ensure doors are closed during radiographic examinations**
> - Sliding mechanism should be sturdy to hold heavy lead lined doors
> - Doors should be easy to slide open/close
> - Sliding mechanisms should be well maintained and kept clean

Walls: materials and lead equivalence

- Walls should be built with material that absorbs radiation, such as 230 mm baked solid clay bricks

- Lead sheet of 2 mm could be sandwiched between other brick types if needed [Figures 3g,h,i]

- Building blocks with openings require use of lead sheets to prevent radiation passing unhindered through the open areas

- Dry walls (wood/chipboard/plywood, etc.) must have lead linings

- Barium plaster at least 6mm of thickness could also be used to cover the walls. Barium has a relatively high atomic number (56) thereby absorbing some radiation

- Walls should be protected up to 2.2 meters from floor level.

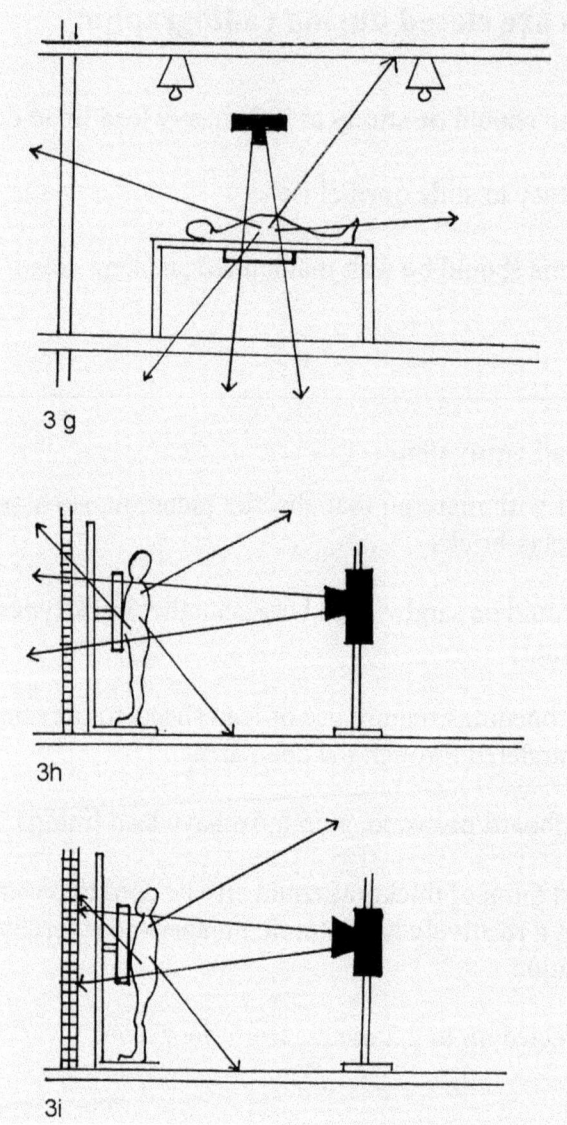

Figures 3g, 3h and 3i: Top diagram (3g) shows primary and scattered radiation passing through ceiling, floor and thin wall as there are no protective barriers in the room. Middle diagram (3h) shows primary radiation passing through a single brick wall which is not an adequate barrier. In the bottom diagram (3i) the room meets international radiation protection requirements as thickness of the walls, floor, and ceiling.

> **Proportions for barium plaster mix**
> - One part coarse barium plaster
> - One part fine barium sulphate
> - One part cement

IMPORTANT REMARK:

The safety of rooms has to be calculated by medical physicists following national laws and regulations. The measurements given above are provided as guidance only.

Ceiling and floors

Ideally, X-ray examination rooms should be on the ground level of a building as this does not require additional protection measures for the floors. If rooms are above ground, the floors should comprise a solid concrete slab of not less than 150 mm thickness. The concrete must be of a high density as recommended by medical physicists (e.g. $2.35 g/cm^3$).

Ceiling slabs must be used in X-ray rooms if the floors above are occupied. Single storey buildings do not require ceiling slabs for radiation protection safety.

Ward radiography

Distance is the most important safety measure to reduce radiation dose to people not undergoing radiographic examination. (see Figure 3b). A minimum of 2 metres from the X-ray tube is usually sufficient provided the X-ray beam is restricted to areas of interest. Beam restriction is discussed in Chapter 5.

Change cubicles

Change cubicles opening into X-ray examination rooms must be lined with 1,5 mm lead sheets, or thicker. Actual thickness would be calculated by a medical physicist. To prevent entrance during radiation exposures, access doors of cubicles should be fitted with locks as a safety measure.

Safety measures in special procedure rooms

The above guidelines for a general X-ray room also apply to special procedure rooms. However, the main aim of this book is to address radiation protection in places with limited resources, where such rooms are rarely available.

Summary

All safety measures available should be implemented to minimize the risks of unnecessary radiation dose to patients, staff and members of the public. These measures include size of room, layout, and materials used. National laws shall be followed. Additionally, there are international guidelines addressing acceptable protective materials used in walls, protective cubicles, floors, and ceilings. Concrete of a specific density and thickness acts as a barrier by absorbing radiation, as do lead linings and lead glass windows. However, the most easily applied important protective measure is increased distance from the source of both primary and secondary scattered radiation.

CHAPTER 4

Radiation protection devices

It is essential that radiation workers be protected when they need to work outside the protective cubicle. There are several essential protective devices, including protective clothing, which should be readily available for use in every X-ray room. These devices are used to protect staff from receiving unnecessary radiation dose from both the primary beam and from scattered radiation. These devices should also be used to shield members of the public from unnecessary dose, for example when a parent holds the arms of a baby during exposure of a chest radiograph.

Lead rubber aprons

As far as reasonably possible, radiation workers such as radiographers, radiological technologists and radiologists should remain in the protected area during exposure. When this is not possible, they should be provided with lead rubber aprons of at least 0.25 mm lead equivalence. If they stand within one meter of the X-ray tube or patient when the unit is operated at tube voltages above 100 kV, they should wear protective lead rubber aprons of at least 0.35 mm lead equivalence. Lead rubber aprons are available as single-sided (protects anterior/front part of body) or double-sided (protects back and front of wearer). If a worker wears a single-sided apron then it is important to always face the source of radiation and not to turn away from the source.

Note that members of the public, who assist during an examination and therefore have to remain inside the examination room during exposure, must be provided with necessary protection devices such as lead rubber aprons and lead rubber gloves.

Care of lead rubber aprons

- To prevent damage to aprons when not in use always hang them up on a sturdy hanger

- Never fold aprons as this could cause cracks in the lead rubber

- Undertake monthly visual inspections of all protective aprons for cracks, splits, rips, tears, etc.

- Aprons suspected to be damaged can be radiographed if in doubt:
  1) Place suspect area of apron on an unexposed loaded cassette and expose to radiation. Using at least 70 kV and 10 – 15 mAs at 100 cm FFD
  2) Process film and inspect for signs of fogging and if noted then withdraw defective apron from use

- Double-sided aprons should be opened fully so that one side at time is checked

- Depending on size and degree of damaged areas, aprons can be repaired. Always re-radiograph a repaired apron to make sure it is suitable for use

- Defective items should not be used.

Protective lead rubber gloves

According to the *ICRP Publication 57*, lead rubber gloves should be at least 0.35 mm lead equivalence. Gloves should be used to protect workers' hands when placed in close proximity or under the primary beam, for example during a barium study. This also applies to any person who is in close proximity to the X-ray beam, such as a parent holding a baby during an X-ray examination [Figure 4a].

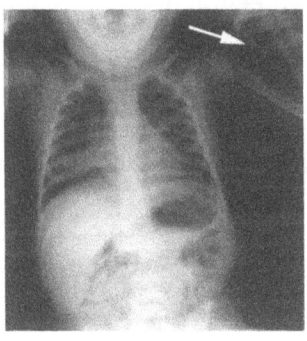

Figure 4a: Arrow shows unprotected hands of helper/parent. This is poor application of radiation protection measures to limit dose to members of the public. It is essential to provide members of the public with lead rubber gloves, or mittens to cover hands within primary beam. Besides the fact that a member of the public received dose to hands, the patient also received unnecessary radiation dose. ALARA was not

implemented as the entire abdomen of the baby was exposed unnecessarily to ionizing radiation. Paediatric chest radiography should always include beam restriction to chest area. This should be done routinely in all paediatric radiography to limit dose to children. This radiograph is a typical example of (i) poor radiographic technique, and (ii) inadequate implementation of radiation protection measures. Also, a black image like this indicates use of too high mAs which contributes unnecessarily to increased patient dose (see Chapter 7).

Lead rubber gloves

- Handle with care to prevent damage

- When not in use, store flat in a safe place within easy reach

- Gloves should be checked monthly for cracks or defective areas. Defective gloves should be withdrawn from use.

Thyroid shields

The thyroid gland is relatively sensitive to ionizing radiation. Therefore, it is recommended to use a radiation protection device whenever possible. There are several types of shields on the market. If not available, a lead rubber apron with a high neckline can be used. Care should be taken when using the shields to ensure they do not get damaged, and they should be stored in a safe place when not in use.

Gonad shields

Whenever possible, gonads should be protected from being exposed to ionizing radiation. When gonads are within the primary beam or within 5 cm of it, some shielding should be used if this can be done without obscuring or excluding information needed for diagnosis.

Gonad shields are of three different types

- Contact shields: these are fairly inexpensive and easy to use as they are made from pieces of lead sheet or lead rubber. Lead gloves can also be used for gonad shielding

- Shadow shields do not come into contact with the patient as they are radio-opaque shields placed between the X-ray tube and the patient

- Shaped contact shields are available for male patients

If a lead rubber apron and lead gloves are beyond repair, parts of these may be cut up and be used as contact gonad shields

Summary

Protective devices/clothing should be used by persons exposed to ionizing radiation related to X-ray examinations of other persons than themselves. Patients should be protected whenever possible provided relevant information is not obscured by the devices.

All protective devices should be inspected on a regular basis to ensure that they are not damaged. The ALARA principle should be applied for every exposure made to patients and this includes use of protective items, whenever possible.

CHAPTER 5

Beam-restricting devices

In accordance with the ALARA principle all possible dose reducing measurements should be employed. Beam-restricting devices play a significant role in dose reduction [Figure 5a].

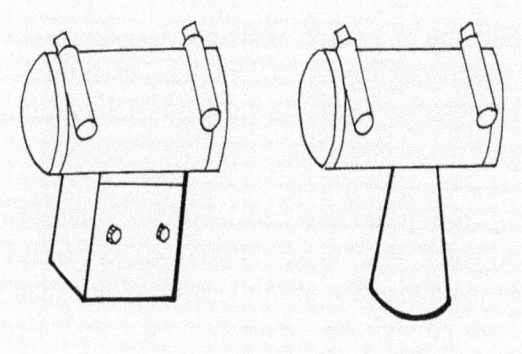

Figure 5a: Line drawing (left) of variable-aperture beam limiting device, i.e. light beam diaphragm (LBD), and (right) removable metal cone.

As scattered radiation adds to dose, all means available should be used to minimize this type of radiation. The larger the area covered by the primary X-ray beam, the greater the amount of scattered radiation produced (Fig. 5b). Apart from contributing to dose, this radiation is in addition to that of the primary beam and adds to film density, which could reduce visualization of important details. In view of this, and to keep scattered radiation as low as possible, the primary beam should be confined to the area of interest for the examination.

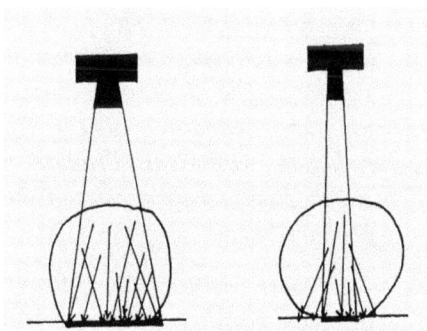

Figure 5b: Diagram (left) of an unrestricted beam; considerable scattered radiation is produced in a thick object. A restricted beam (right diagram) results in production of less scattered radiation; dose to surrounding parts of the object is reduced as no longer in the path of the primary beam.

There are several devices which can be attached to the X-ray tube to restrict the X-ray field size. Beam-restricting devices may be detachable cones or permanent fixtures. Examples of beam-restricting devices include:

- **Aperture diaphragms** are sheets of lead with circular, square, or rectangular openings. These devices are inserted into the X-ray beam near the tube window and are usually used together with a cone or a variable-aperture beam-limiting device.

- **Cones** are metal tubes that come in a range of shapes and sizes. The length of a cone and size of an opening will affect the size of the X-ray field. Cones are usually detachable to allow use of a selection of different sizes and shapes.

- **Variable-aperture** beam-limiting devices consist of lead plates or shutters, which can be adjusted to change the size of the beam, usually by turning indicator knobs. Some modern units are made with a beam-limiting device whose shutters are controlled automatically by sensors adjusting the field size to the size of the cassette (image receptor). These devices contain "cross hairs", a light source, and a mirror to project the light onto the centre point of the patient and to visually indicate outline of X-ray field in accordance with the open shutter size.

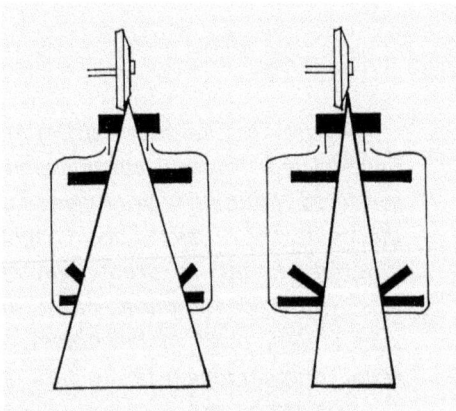

Figure 5c: Example of a variable-aperture, multiple-shutter beam limiting device. Note diagram on the right: size of the primary beam is reduced due to smaller size field. The diaphragm closest to tube window results in reducing off-focus radiation which does not form part of the primary beam.

Apart from the above devices, additional items could

be used to reduce scattered radiation reaching the film. For example, lead blockers can be used to divide a cassette when more than one exposure is made on a single film. Lead blockers reduce unnecessary exposure to both the exposed and unexposed sections of the film thus limiting the chances of film fog and thereby loss of image quality.

Use of lead blockers to improve image quality

- Place a lead blocker on the couch when doing lateral projections of thoracic spine, lumbar spine, sacrum and coccyx spine;

- Position the lead blocker next to the patient's back to absorb radiation that will not pass through the spine. The lead blocker is placed almost in contact with the posterior skin surface of the patient's back;

- Defective lead aprons could be cut into a range of sizes if lead blockers are not available.

Quality assurance tests of beam-restricting devices

It is good radiographic practice to limit the area of radiation so that the resulting image has collimated edges on all four sides of the film. Thus, proper beam-restricting measures (collimation) applied for chest radiographs means minimized or totally excluded radiation to lens of the eyes, as well as other organs not targeted. An example of poor usage of beam-limiting devices is shown in Figure 4a *(page 28)*; the patient received unnecessary radiation dose to the abdomen whilst undergoing chest radiography.

Working tips

- Beam-restricting devices should be checked regularly to ensure they are functioning optimally;

- From time to time during the course of the day, it is strongly recommended to bring the tube to rest on the couch (table top) to check that all four edges of the square light beam diaphragm are in contact with the table/couch top when using a straight vertical beam. Sometimes the tube, due to constant use, moves slightly from its position hence the central ray is not at right angles to the couch.

The same check should be performed with cones when inserted in front of a tube window. Cones sometimes become bent when dropped or bumped and cause irregular and insufficient coning.

Collimator-beam alignment test

- Even if there is no obvious problem, the device should be checked at least every six months to assess proper alignment of the beam-restricting device and primary beam;

- Poor alignment causes sub-optimal positioning of patients as it may be difficult to centre accurately. Thus, the area of interest may not be included on the image, or additional and not wanted areas may be included. In both examples this would lead to increased radiation dose to the patient.

Method for checking the collimator-beam alignment

- Place metal coins or paper clips on a loaded 24 x 30 cm cassette [Figure 5d];

- Set the beam-restricting device (collimator) at 20 x 20 cm and expose the film at 100 cm using 60 kV and 4 – 8 mAs. Process the film and check that the alignment conforms to international performance criteria which allow +/- 2% difference of FFD [Figure 5e];

- Measure the outer edges of the image and the outer edges of the metal coins/paper clips;

- Should the difference exceed the acceptable range of variation, then repair is needed.

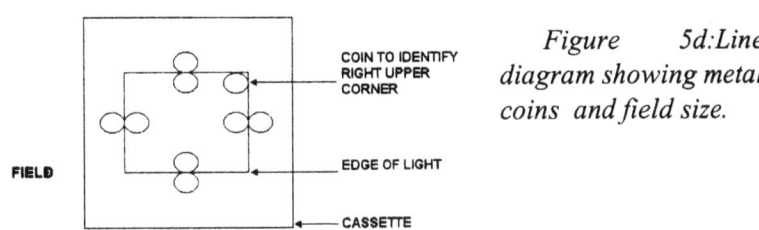

Figure 5d: Line diagram showing metal coins and field size.

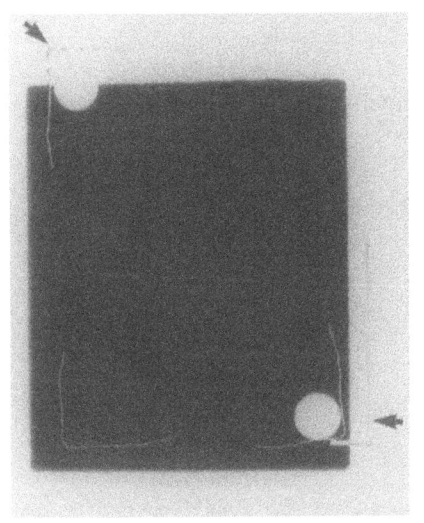

Figure 5e: Arrows indicate light beam edges. Area of film blackening does not overlap light edges. Measurements of both should be taken to determine whether the radiation field size exceeds light field by more than 2% based on the actual FFD.

Test to check alignment of the centre of the X-ray beam

Method

- Place an unexposed loaded cassette in the centre of the bucky tray and centre tube to the cassette;

- Move tube to 100 cm FFD to bucky tray. Reduce the longitudinal collimators to a thin slit (e.g. 0.5 cm). Close the lateral collimators. Expose using 60 kV and 4 – 8 mAs;

- Do not remove the cassette;

- Close the slit collimators and open the lateral ones to collimate laterally to a thin slit (e.g. 0.5 cm). Expose the film again;

- Process the film and inspect the image;

- Bend the film in half and check that the exposed 'cross' is in the centre of film [Figure 5 f];

- This simple test can be used to check alignment of the central ray when performing non-bucky radiography. Remember to increase FFD to measure 100 cm to top of table/cassette;

- A deviation of 1 cm either side of the centre is acceptable, i.e. total of 2 cm;

- Deviations greater than the acceptable range should be corrected to reduce repeat radiographs.

Figure 5f: Arrows indicate middle of film. Cross and film centre are not aligned.

Use of compression to reduce thickness of patient

A further dose reducing measurement is to compress the area of interest, i.e., to reduce thickness of a patient by compression to minimize production of scattered radiation. The thinner the area irradiated is, the less scattered radiation is produced [Figure 5g].

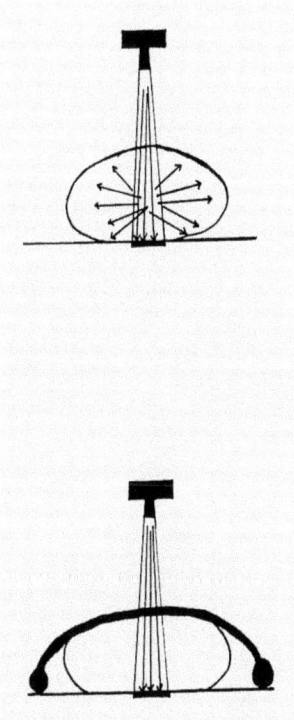

Figure 5g: Example of cross section of abdomen of an obese patient; scattered radiation is in all directions (top). Less scattered radiation produced when overall volume of abdomen is reduced by means of a compression band (bottom).

Summary

Beam–restricting devices play an important role in radiation protection. A device, which is not checked regularly could result in degradation of the image quality and possibly re-exposing a patient. Regular performance of simple quality assurance tests is in accordance with ALARA. The practical implications of the use of beam-restricting devices on exposure factor selection are covered in Chapter 7.

CHAPTER 6

Scattered radiation: role of grids

Scattered radiation should be minimized (controlled) in all radiographic examinations. The use of beam-restricting devices 'control' scattered radiation before the primary beam enters the patient. Reduction of scattered radiation from the patient reaching the image receptor/film is achieved by using a grid. Not only do grids 'clean-up' scattered radiation for overall improvement of the image, but they also reduce the scattered radiation continuing in forward directions after having penetrated the patient. The intensity of the scattered radiation depends on the kV selected; this is discussed in-depth in Chapter 7.

Grid design

A grid consists of alternating strips of lead to absorb scattered radiation, and a spacer material, such as transparent fibre, that does not absorb ionizing radiation. The strips and spacers are encased in a protective firm cover. The cover must be sturdy to ensure the grid strips are aligned correctly to the rays of the primary beam [Figure 6a]. The spacer material is sandwiched between the lead strips and this allows most of the primary rays to pass through onto the film for image formation. The oblique rays of the scattered radiation are absorbed by the lead strips because they are usually at an angle to the lead strips.

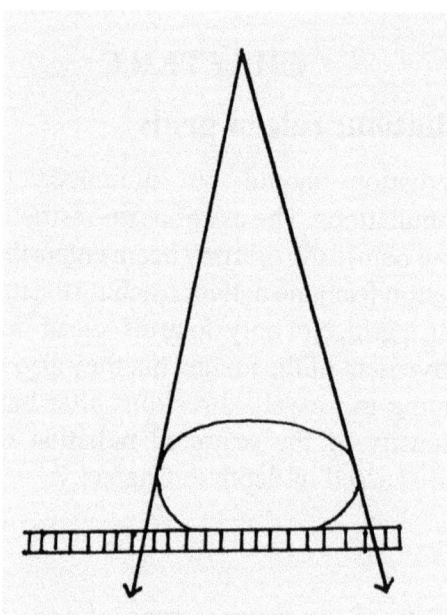

Figure 6a: Diagram to show spaces between lead-strips allowing the primary beam to pass through without being absorbed.

Grid ratio

A range of grids is available. Some are more efficient than others. The most important factor is the *ratio* of a grid; this is the relation between the depth/height of the lead strips and the width of the transparent spacers sandwiched between each strip [Figure 6b]. For example, if the height of the lead strip is 8 times the width of the inter-space, then the grid ratio is 8:1. If the height is 10 times the width of the inter-space, then the ratio is 10:1. The greater the grid ratio is, the more efficient the grid is in absorbing scattered radiation. However, some primary rays also get absorbed by a grid. Manufacturers clearly mark the ratio on at least one side of every grid and this is important for selecting the correct exposure factors as discussed in Chapter 7.

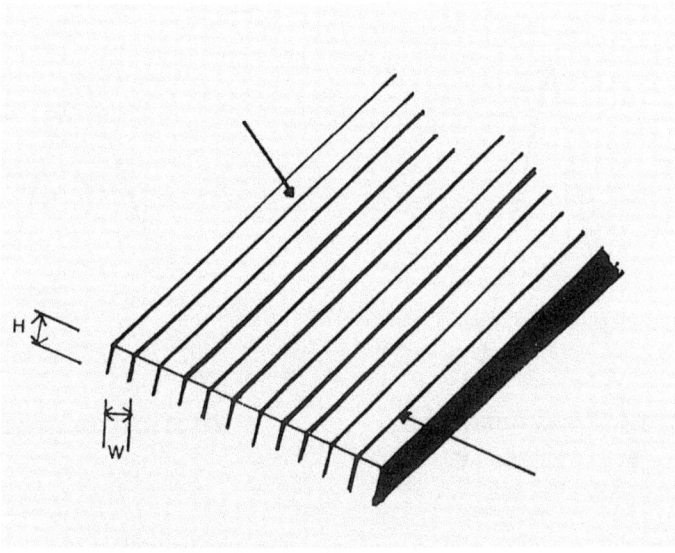

Figure 6b: Diagram showing width and height of grid strips. Arrow on left indicates inter-space. Arrow on right indicates lead-strip.

Parallel and focused grids

The most common types of grids for general radiography are "parallel" and "focused".

- Parallel grids consist of parallel lead strips [Figure 6c] resulting in absorption of some of the oblique rays of the primary beam as well as the scattered radiation

- Focused grids consists of lead strips that are progressively angled to accommodate the oblique rays of the primary beam. The angled strips converge to a point; the distance from the point to the grid is its focal distance, or length. The focal length of a grid is always clearly marked on it. Thus, a focal length of 100 cm means that the grid is effective in "cleaning up" scattered radiation without affecting the primary beam significantly at an FFD of 100 cm. If the FFD were shorter or longer then indicated as the focal length of the grid, then absorption of the primary rays would increase. This is called "cut-off" [Figures 6 d, e ,f].

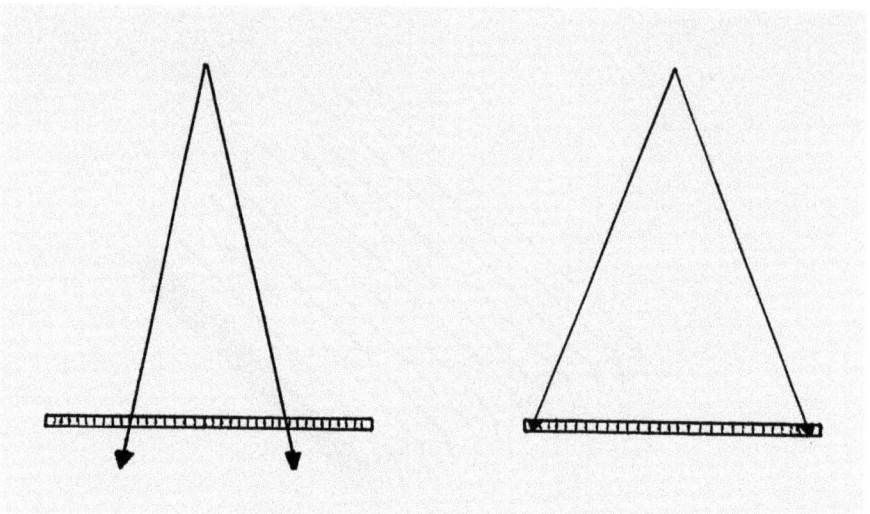

Figure 6c: Example of beam passing through spaces in a parallel grid. Note the absorption of outer oblique rays in right diagram.

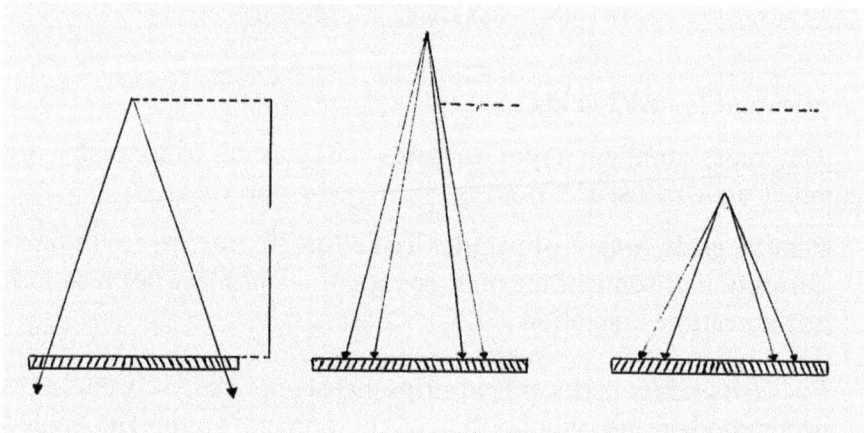

Figures 6 d, e, & f: (left) rays pass through inter-spaces of focused grid at correct focal distance. Increased distance (middle) shows some grid cut-off. Right diagram shows cut-off when distance decreased. To minimize retakes it is important to use correct FFD for focused grids to prevent grid cut-off.

Grid cut-off

Ideally, the central ray of the primary beam should coincide with the centre of the grid so that the primary rays intersect the grid perpendicularly. If this is not the case, a certain amount of "cut-off" occurs causing a progressive decrease in the transmitted X-ray intensity towards the outer edges of the grid [Figure 6 h].

Cut-off can result when the tube is tilted laterally across the lead strips [Figures 6 i, j]. This is a common problem when the cassette is not placed on a flat sturdy surface.

Parallel grids tend to produce cut-off when used at long distances from the X-ray tube. Care should be taken when placing a parallel grid to avoid cut-off.

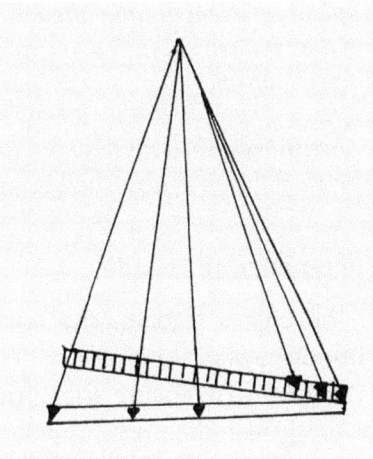

Figure 6h: Line diagram showing grid cut-off when central ray is not perpendicular to the grid. This problem occurs when using a focused grid in ward radiography. For example, cassette and grid placed under patient lying on a soft mattress results in non-alignment of central ray and lead-strips of grid.

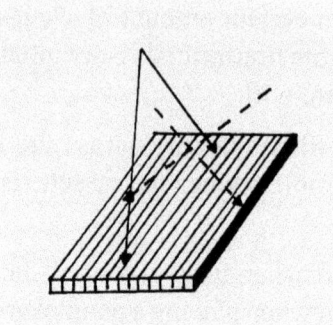 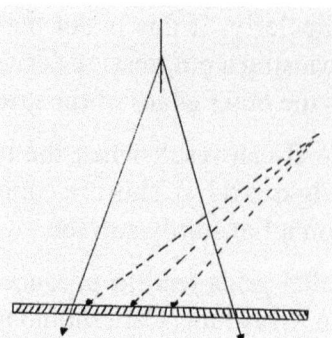

Figure 6i (left): Typical problem of tube tilted laterally across lead strips resulting in grid cut-off thus patient subjected to extra dose as radiograph has to be repeated without grid cut-off.

Figure 6j (right): Central ray tilted across lead strips causes cut-off. Note: When this occurs the primary beam is absorbed by the lead strips.

Focused grid: importance of tube-side

As indicated in the above figures, the lead strips are angled to accommodate the oblique rays of the primary beam to pass through the spacers and reach the film. To ensure that grids are correctly used, manufacturers mark the grid indicating tube side. If a focused grid is inadvertently placed upside down, most of the primary rays will be absorbed by the lead strips [Figures 6 k and l].

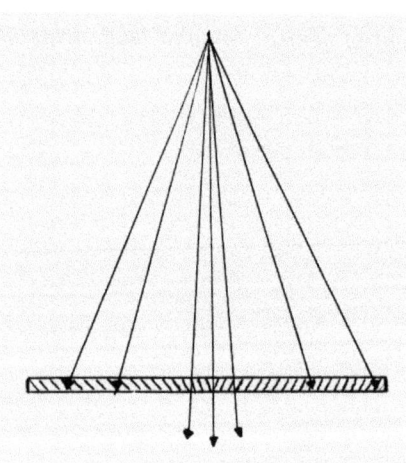

Figure 6 k: Line diagram showing absorption of primary beam because focused grid is incorrectly placed, i.e. upside down resulting in tube side of grid facing the film/receptor.

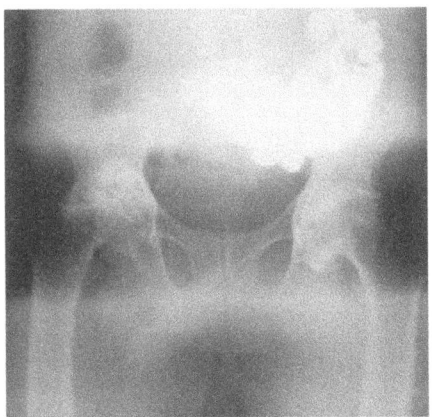

Figure 6 l: AP pelvis showing extensive grid cut-off because focused stationary grid was put up-side down by mistake causing insufficient implementation of ALARA as the radiograph had to be repeated.

Stationary grids

When performing out of bucky tray work, a stationary grid is used. The grid is placed on top of the cassette and positioned under the patient's body part being examined. When a stationary grid is used, grid lines may be evident on the resultant film [Figure 6 m]. The thinner the grid the less obvious the grid lines. Stationary grids can be used with cassette holder trays under the examination/couch table. For example, fine stationary grids may be supplied in skull units that do not include moving grid mechanisms.

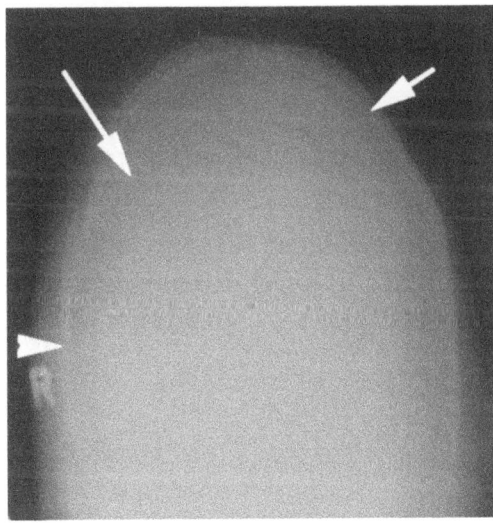

Figure 6 m: Arrows indicate stationary grid lines.

Moving grids

These grids move during the exposure to blur out the grid lines thereby improving visualization of image detail. Some grids move from one side to the other, whilst other move constantly from side to side and are called reciprocating mechanisms. When using moving grids, the exposure time must not be too short to avoid producing a striped pattern [Figure 6 n]. As with stationary grids, the "tube side" of a moving grid must face the tube [Figure 6 o].

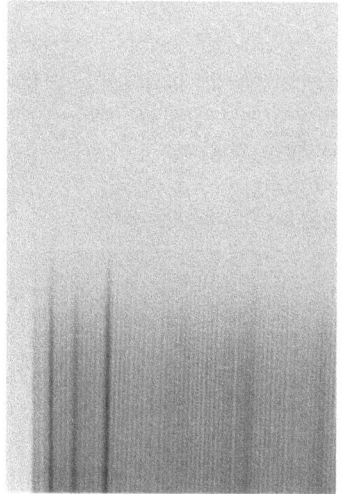
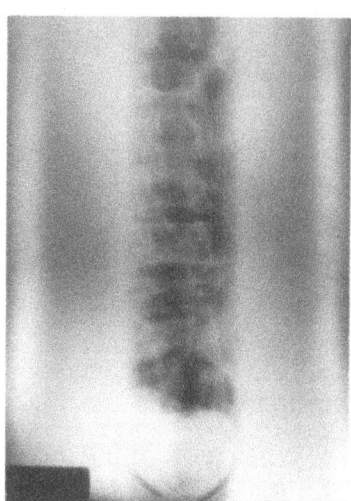

Figure 6 n (left): Too short exposure used together with a moving grid, results in thick black grid lines on a part of the image.

Figure 6 o (right): Radiograph AP abdomen showing only centre area of film density. Here, a focused moving grid had been placed upside down causing absorption of oblique primary rays.

Scatter clean-up: role of grids

As thick body parts produce more scattered radiation [Figures 6 p, q], grids are normally used to reduce this.

When examining (i) infants and small children, (ii) upper extremities of adults to mid region of humerus, and (iii) lower extremities to mid-thigh region of humerus on thin patients, grids are normally not needed, nor recommended.

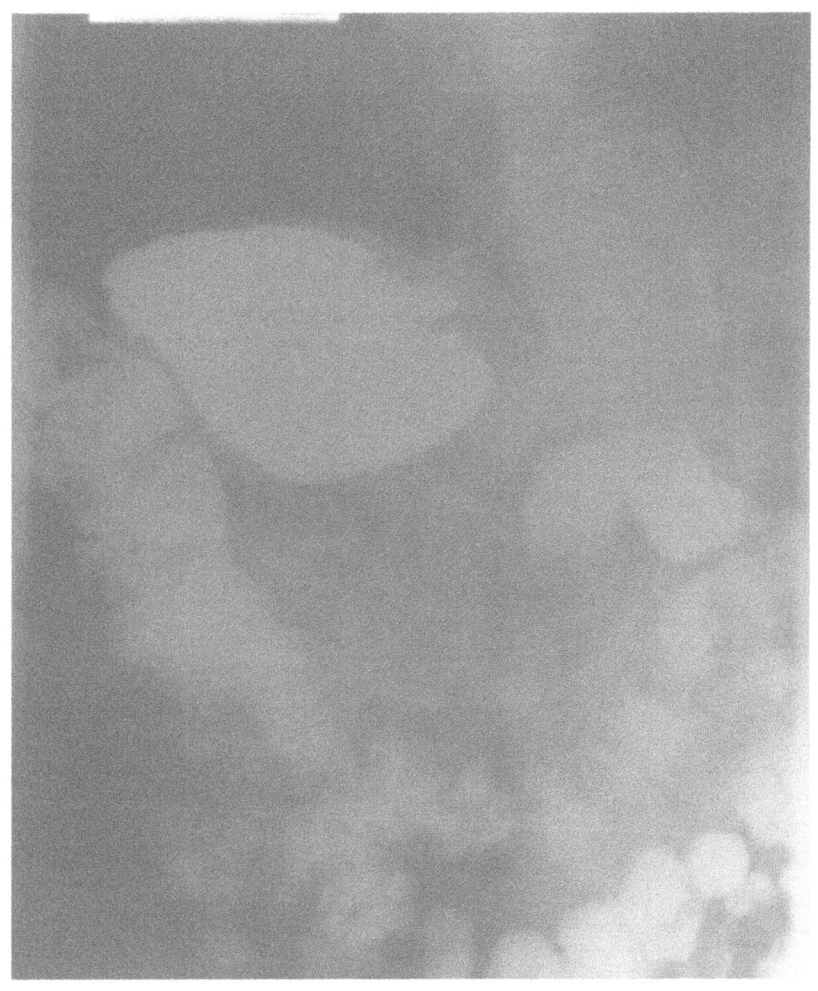

Figure 6 p: Fogged barium meal radiograph with very low subject contrast, i.e., differences in film densities. Image quality is poor because no grid was used; the technician forgot to replace it after a routine service. Scattered radiation caused 'fogging' of film.

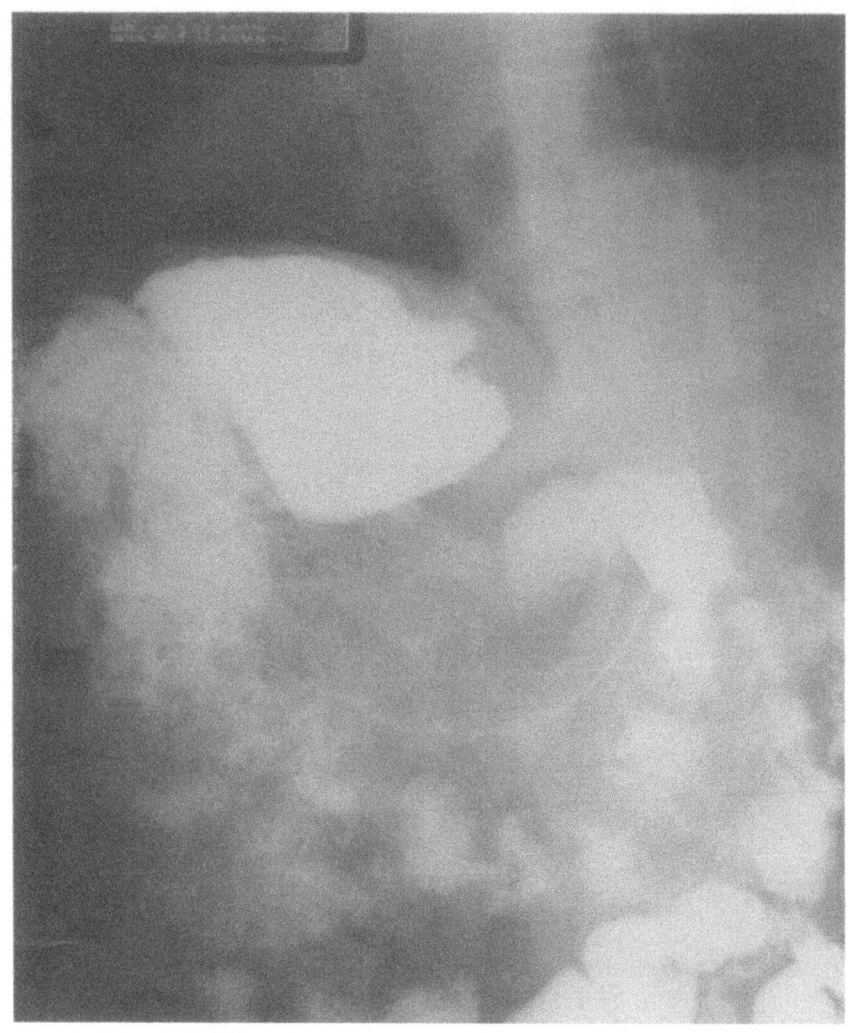

Figure 6q: Barium meal repeated after the grid was replaced. Note radiographic contrast greatly improved; scattered radiation absorbed by the grid.

Care and maintenance of grids

It is advisable to test a new grid for cracks before accepting it from the supplier. Also check that the grid performs according to its specifications when used at correct focal length. Thus, a 150 cm focused grid should not produce cut-off when used at this FFD.

Grids should be handled with care. When not in use, they should be stored in upright positions to avoid possible damage. Placing heavy objects on a grid could cause it to bend or crack. Keep sharp instruments away from grids as these could pierce the lead strips. Grids should not be dropped as this could result in cracks or bends resulting in non-alignment.

To prevent film artefacts, the outer casing of a grid should be cleaned using a damp cloth. Testing of grids should be included in the quality assurance programme of the department to ensure the grids function optimally. A bent grid results in grid cut-off and may require the patient to be unnecessarily re-exposed to ionizing radiation.

Quality assurance: grid tests

Grids should be checked at regular intervals to evaluate their efficiency. A poorly functioning grid does not benefit the patient.

Method

- Inspect the grid for bends or cracks and if found, take a radiograph of the grid;

- Place the grid on top of an unexposed, loaded cassette of the same size or slightly bigger than the grid. The grid must be placed with tube side facing the tube. Cassette must be on a flat surface so that the central ray of the X-ray beam will be perpendicular to the grid;

- Select the relevant FFD to match that of the focal length of the grid;

- Centre the tube to the centre of the grid. Central ray must not be tilted;

- Expose using a low exposure, e.g. exposure factors for adult finger or hand;

- Process the film and view the radiograph [Figures 6 r and s];

- Defective grids should not be used as they could contribute to production of poor image quality.

Figure 6 r (above) and Figure 6 s (below): Examples of malfunctioning grids. A radiograph of an efficient grid should show thin grid lines with overall even film density.

Grid factors

Since grids absorb scattered radiation contributing to the overall density of the film the exposure has to be adjusted. Some of the primary beam is also absorbed by the grid causing reduced density of the film. Consequently,

- the overall beam intensity (mAs) must be increased to compensate for loss of film density when a grid is used;

- the grid ratio determines how much the mAs has to be increased. Half the grid ratio, called the grid factor, is the amount by which the mAs must be increased;

- an 8:1 grid ratio has a grid factor of 4, whereas a grid ratio of 12:1 has a grid factor of 6. When using the latter, the mAs would have to be increased 6 times compared to non-grid exposure factors; a grid factor of 4 would require a 4 times increase in mAs.

In other words the more efficient a grid is in absorbing scattered radiation, the more the mAs has to be increased to compensate for loss of film blackening (i.e., density). This aspect of the grids is covered in Chapter 7.

Summary

When examining thick parts of the body, it is essential to minimize the scattered radiation for overall improvement of the image quality. Grids must be aligned correctly to the central ray to prevent grid cut-off. Grids should be checked regularly as a part of a quality assurance programme to reduce possible need to re-expose patients to ionizing radiation. When using grids, the mAs is increased when compared to non-grid exposure factors to compensate for loss of overall film density produced by both the primary beam and scattered radiation. Grids are costly items and should be handled with care.

CHAPTER 7

Radiographic technique, exposure factors, and quality assurance tests

A radiograph consists of a range of film densities called radiographic contrast. The most important factors influencing film density, or contrast are:

- The tube current, i.e. milliamperage (mA) discussed in Chapter 2 ;

- The distance between the tube and the patient and film, i.e. application of the inverse square law;

- The quality of the beam, i.e. kilovoltage (kV), and voltage waveform (output of generator) mentioned in Chapter 2.

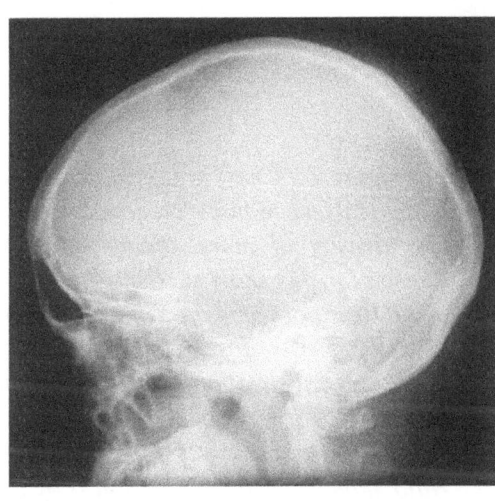

Figure 7a: Lateral skull radiograph with high contrast (short scale); insufficient kV used thus dense skull bones were not penetrated.

Due to the unique anatomical structures of each patient, the contrast of the resulting images is called subject contrast, and can be altered by manipulating various factors, such as exposure factors, use of beam limiting devices, use of compression, and introduction of a contrast medium. The reason for altering these factors is to comply with the ALARA principles.

An image with mostly white and black areas (i.e. large differences in film density) is defined to be of *short scale contrast*; [Figure 7a]. If an image has shades of grey, it will have a *long scale contrast* [Figures 7b and c]. These different shades of contrast of the image depend on the selection of kV and the output of the generator.

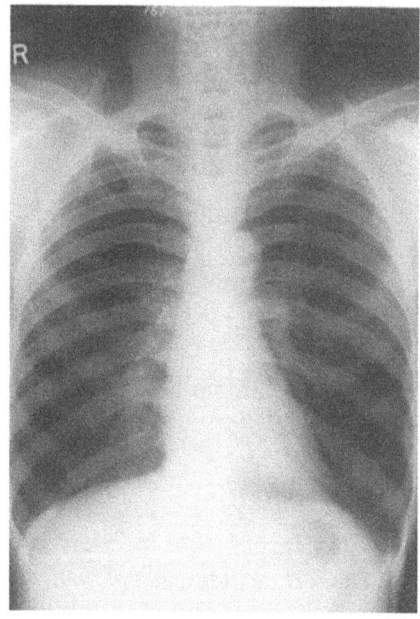

Figure 7 b: High contrast/short scale chest radiograph obtained using 46 kV and 25 mAs, no grid. Low kV results in high absorbed dose. Image does not have range of film densities.

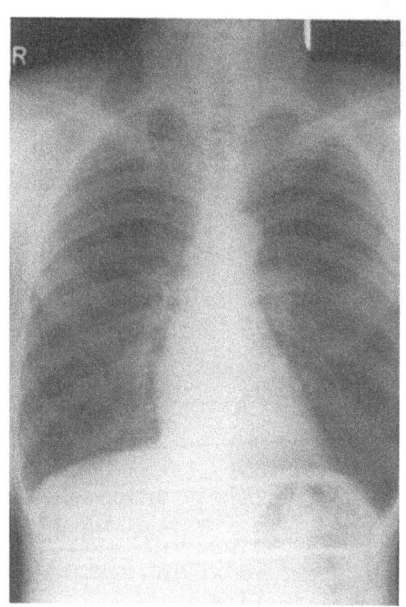

Figure 7c: Chest radiograph of same patient taken 6 months later. Note overall film density of more shades of grey. Reason for this short scale contrast film could be exposure factors changed to 85 kV and 2.5 mAs; dose to patient reduced due to less absorption of primary beam and use of low mAs. Conversion of factors based on rule of thumb: kV/mAs ratio, namely ↑10 kV then ÷ mAs by 2. A more scientific method: ↑ kV by 15% then ÷ mAs by 2.

The choice of required radiographic projections should be based on a correct assessment of the clinical indications and the expected yield from each examination and its contribution to further medical management of the patient. If a patient has recent previous films, these should be viewed before embarking on another examination.

Before commencement of any radiographic examination on a female patient of child bearing age, it is essential to determine whether she might be pregnant (see Chapter 8).

Patient positioning

To reduce the need for repeated examinations, care should be taken when positioning the patient. Immobilization devices, e.g. sandbags, compression bands, head clamps, should be used to reduce risk of patient movement during the exposure.

When examining the upper limbs of a patient, it is important to make sure that the patient's gonads are not within the primary beam [Figure 7d]. Beam restricting devices, as discussed in Chapter 5, should always be used to limit dose to the region of interest only. It is important to remember that the field size of the beam entering a patient produces a larger field size on the film because the beam diverges (fans out). So when considering the size of beam restriction (coning) to minimize dose, it is preferable to select size of required collimation before positioning a patient [Figure 7 e].

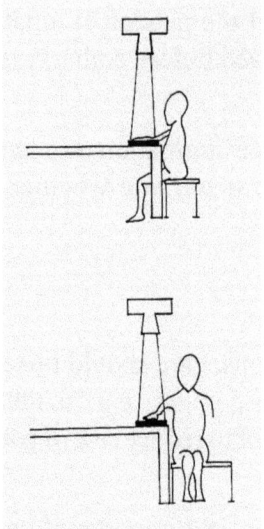
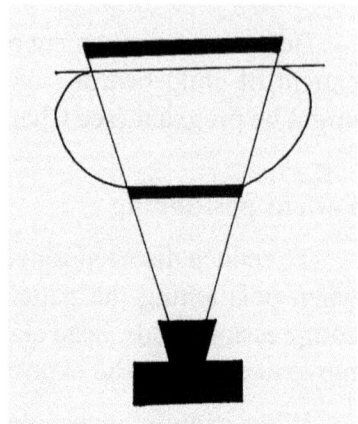

Figure 7d (left): Top diagram: Poor radiographic practices not in keeping with ALARA as the patient's gonads are in the primary beam. This example of incorrect patient positioning is not uncommon as some radiographers tend to forget basic radiation protection measures. Bottom diagram shows good implementation of ALARA.

Figure 7 e (right): Note field sizes. The one closest to tube is smaller than the one closest to film/image receptor.

Skull projections requiring caudal tilt of the tube, i.e. towards the feet of the patient, should be done with a very good restriction of the beam to minimize dose to thyroid. For example, a fronto-occipital 30 degree caudal tilt (Townes view) requires the patient to be facing the tube. If not well collimated, it may result in unnecessary radiation dose to the thyroid gland.

Skull radiography includes the orbits. Therefore care should be taken to reduce dose to eyes as mentioned in Chapter 2. Excessive radiation dose to the eye lenses could result in cataracts.

Chest radiography should whenever possible, be done PA to reduce dose especially to patient's eyes, thyroid, and breast tissues. Beam should be reduced to only expose the chest area (i.e. area of interest). It is essential to produce chest images taken on full inspiration to ensure that the entire lung fields are visualized [Figure 7 f].

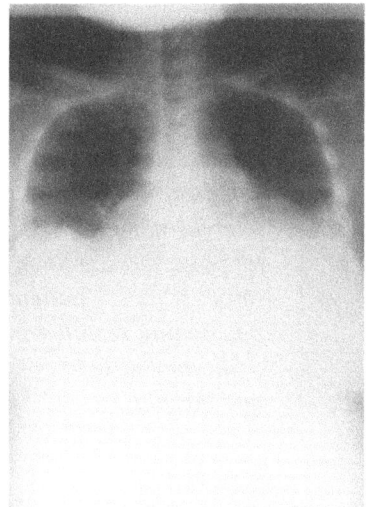

Figure 7 f: AP chest radiograph with minimal visualization of entire lung-fields because expiration film. Full inspiration chest radiograph is considered diagnostic for pattern recognition.

Most examinations require a minimum of two projections at right angles, such as AP and lateral. The patient must be correctly positioned to ensure the projections are at right angles. For example, when performing an AP radiograph of an ankle, it is important to check that the lateral and medial malleoli are equidistant to the film. This often requires slight medial rotation of the ankle and foot. Most patients find this position difficult to maintain. Thus, it is recommended to use sandbags to prevent the patient moving. Soft pads should be placed against the patient's ankle and supported by a sandbag.

Radiation protection includes the use of all methods to reduce dose to patient, staff, public and the environment. There are numerous publications on radiographic techniques pertaining to correct patient positioning, direction and angle of ray, and suggestions for use of immobilization devices [Figure 7 h]. One of them is the WHO Manual of Diagnostic Imaging: Radiographic Technique and Projections (ISBN 92-4-154608-5).

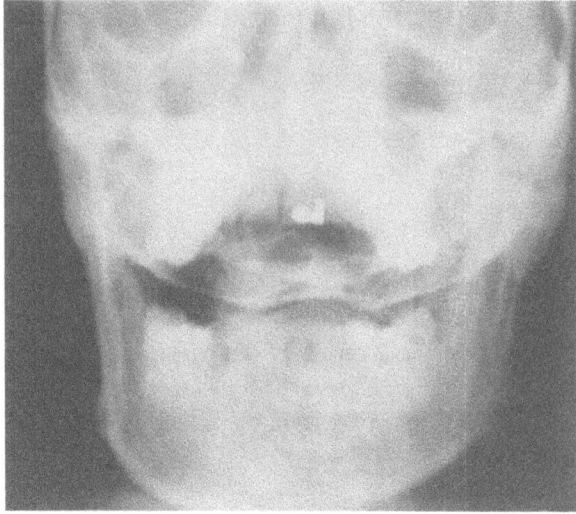

Figure 7 h: Inadequate open-mouth peg view: C1/C2 not visualized due to overlying dense skull bones. Film repeated to demonstrate area of interest, namely C1/C2 for pattern recognition. Poor patient positioning resulted in additional dose to eyes and thyroid.

Selection of kV

A change in kV results in a change in the penetrating power of the X-rays and the overall intensity of the beam. When kV is increased, shorter X-ray wavelengths with greater penetrating power are produced. Penetrating wavelengths have the ability to pass through dense structures, such as dense bones or dense plaster of Paris [Figure 7 i]. Because the beam penetrates dense structures, not much of the ionizing radiation is absorbed in the patient. Also, absorption of scattered radiation is reduced when the kV is increased [Figure 7 j]; the use of a penetrating beam is one method of reducing absorbed dose to a patient. However, this does not mean that all images should be obtained using high kV techniques as it may be necessary to visualize soft tissues and/or surrounding structures [Figure 7 k]. Soft tissue visualization requires use of low kV techniques [Figure 7 l].

Scattered radiation in high kV is energetic. Thus, the angle of the rays is not much different to those of the primary beam. Because high kV scattered radiation is not deflected very much, it means that it follows the path of the primary beam in a forward direction adding to film density. Use of relatively high grid ratios, e.g. 12:1 or higher if available, improves image quality as the scattered radiation in high kV techniques would be more efficiently absorbed.

Figures 7 m-n are examples of low kV *versus* higher kV imaging in radiography of knees; note there should be good detail of the patella on AP knee radiographs and this is usually achieved by adequate penetration of the primary beam [Figure 7 o].

Selection of mAs

The amount of mAs used to produce an image has a direct bearing on radiation dose to patient [Figures 7 p, q and r]. Use of a high mAs setting means increased dose to patient because the intensity of the beam is increased. The most important factors affecting mAs selections are:

- Focus film distance (FFD);

- Output of generator/capability of unit;

- Speed of film;

- Size, thickness, and type of phosphors in intensifying screens;

- Use of grid;

- Degree of collimation.

- Relationship to kV used;

- Size of focus.

Due to the inverse square law an increased FFD requires increased mAs to produce a comparable image based. For example:

- Increasing FFD two times, such as from 90 cm to 180 cm requires a four- times higher mAs.

Selection of mAs depends on output of generator to produce comparable images. When using a 3-phase generator instead of a single-phase one, a decrease in mAs is required. Thus, radiation dose to patient is reduced.

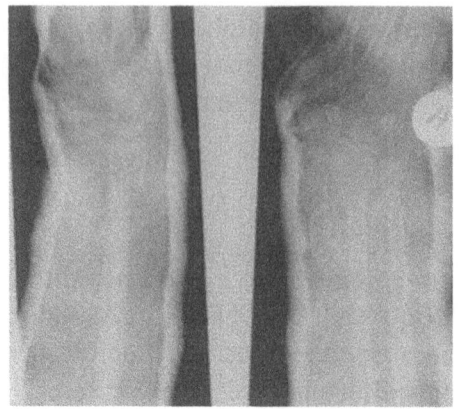

Figure 7 i: Inadequate penetration of wrist in plaster of Paris (POP). White/black film indicates that kV was not increased sufficiently to penetrate the POP. Patient received additional, unnecessary radiation dose as the examination had to be repeated.

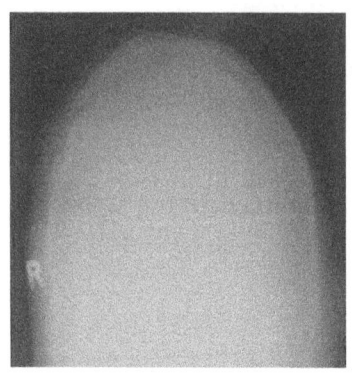

Figure 7 k: Low kV selected to demonstrate fracture of skull.

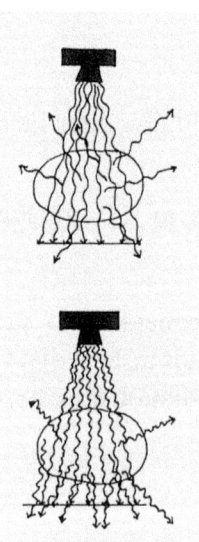

Figure 7 j: Low kV (top diagram) produces more scattered radiation than high kV (bottom diagram).

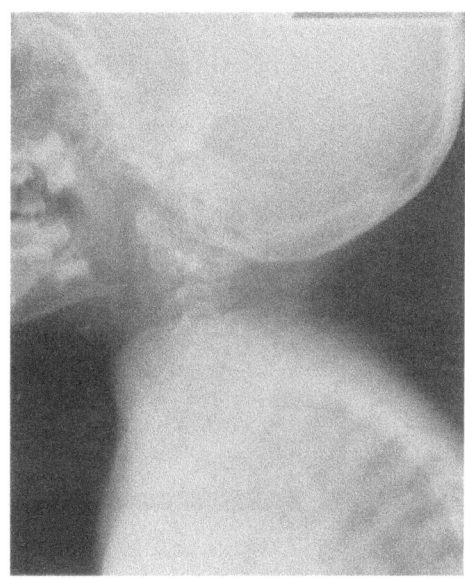

Figure 7 l: Low kV technique for soft tissue lateral neck examination on a child. Lack of evidence of beam restriction. Note artefacts (ear-rings). Retake done because not true lateral. Child received additional radiation dose to eyes and thyroid.

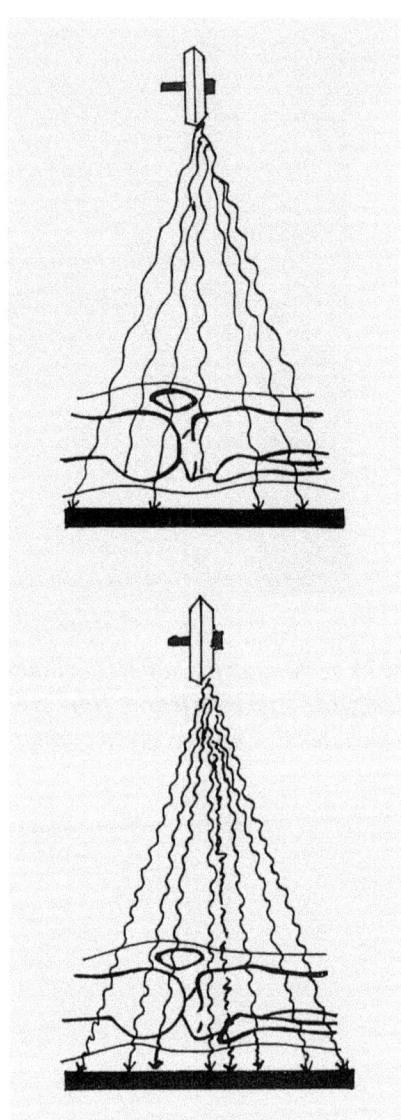

Figures 7 m: Line diagrams to show effect of low versus high kV: Low kV (top) results in high absorption of ionizing radiation whereas high kV (bottom) results in less absorption, or in other words, that more ionizing radiation reaches the film after having penetrated the tissue examined thereby causing a good visualization of the structures as required in pattern recognition.

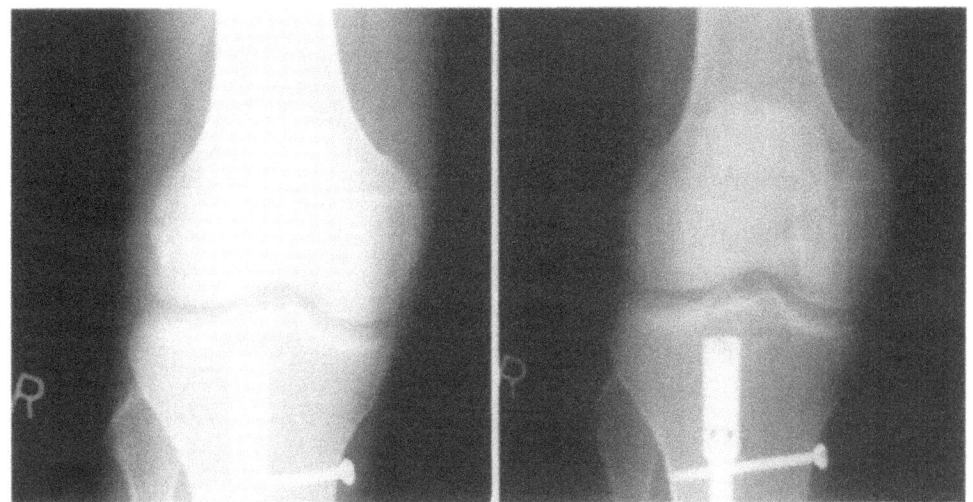

Figure 7 n: AP knee (left) obtained using 55 kV and 9 mAs thus poor penetration of dense knee structures. Good penetration of knee structures (right) on follow-up images a few weeks later: exposure factors 70 kV and 3 mAs thus dose to patient reduced.

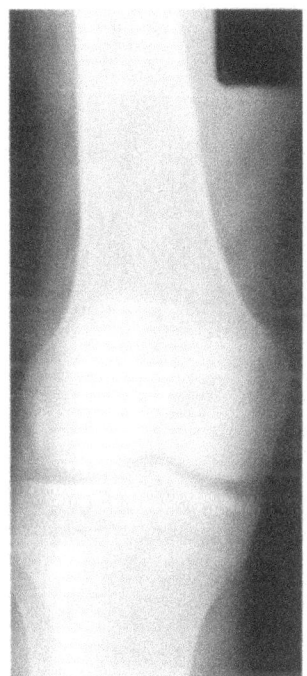

Figure 7 o: Poor patient positioning: entire knee architecture not included. Lack of penetration of knee joint due to low kV selection.

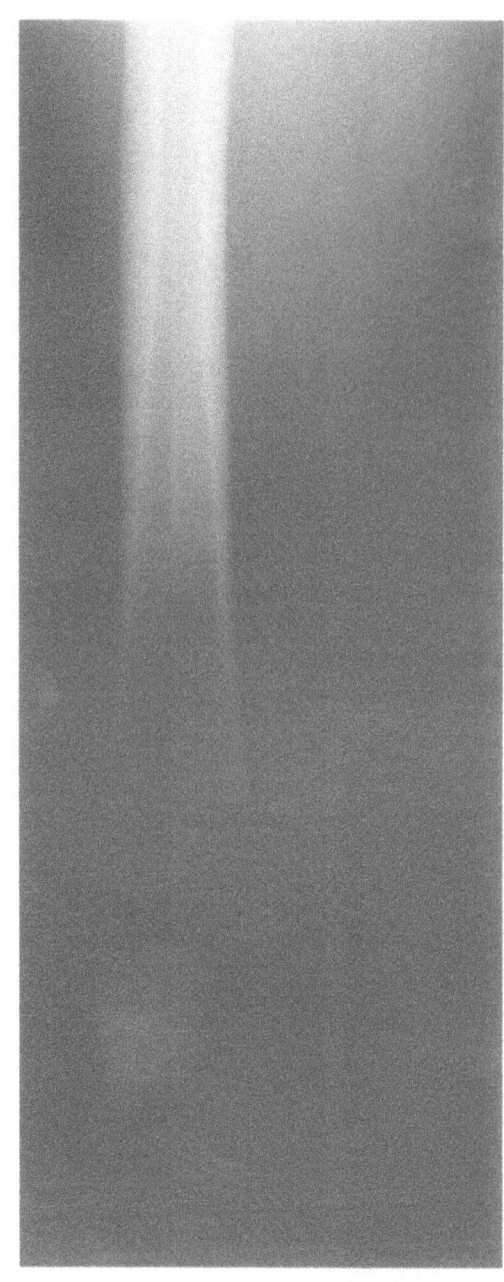

Figure 7 p: Very black film of distal femur due to high mAs. Patient subjected to unnecessary dose; film retaken with reduction of mAs.

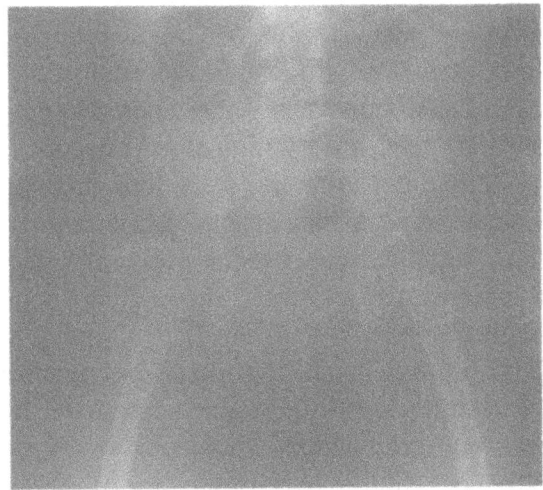

Figure 7 q: Implementation of ALARA not evident in black AP pelvis of child. Radiograph grossly over–exposed as mAs too high plus beam not restricted. Child's gonads in primary beam, which means dose to reproductive organs. Radiograph repeated.

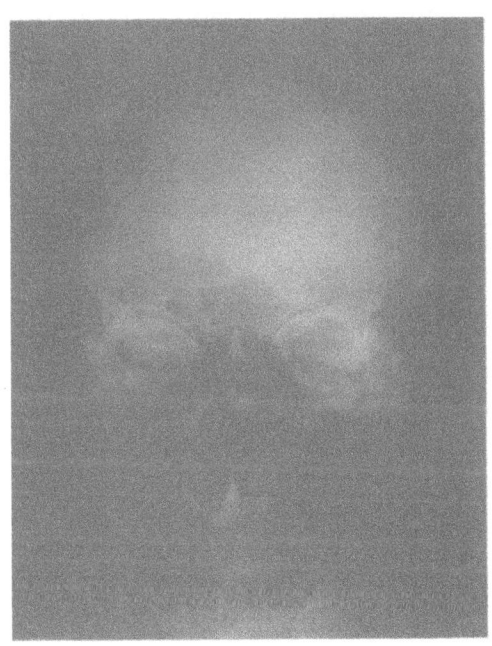

Figure 7 r: Black skull projection as mAs too high. Poor radiographic technique as patient's skull rotated. No evidence of beam restriction thus unnecessary dose to patient's eyes and thyroid. Poor implementation of ALARA.

Fast films need less mAs in order to produce comparable film blackening compared to slow films. The same applies to intensifying screens. Fast screens require less mAs slow ones. Thus, when changing from a 200 speed system to a 400 speed system original mAs needed is reduced by 50%, meaning that the patient receives less ionizing radiation. Selection of mAs is determined by the intensification factor of the intensifying screens. Thus, large phosphors require less mAs than smaller ones to obtain the same film blackening. Similarly, a single intensifying screen would require more mAs than having two screens in a cassette. Slow screen systems produce good image detail, but as more mAs is needed for film blackening, dose to patient is higher. The selection of specific film-screen combinations should be based on information required for pattern recognition purposes.

It would be of no benefit to the patient if a system is used that does not have good film-screen contact as information would be lost [Figure 7s]. Testing film-screen contact should be included in the department's QA programme to avoid subjecting patients to unnecessary ionizing radiation.

Figure 7 s: Arrows indicate poor film/screen contact.

When using a grid to minimize scattered radiation, the mAs must be increased based on the grid factor as discussed in Chapter 6. Limiting the beam to the area of interest requires an increase in mAs as discussed in Chapter 5. Figures 7 t and u clearly show how beam restriction without an increase in mAs reduces film density. Suggested increases in mAs when reducing film size:

- 35 x 43 cm field size to 25 x 30 cm: ↑mAs approximately 15%.

- 35 x 43 cm field size to 20 x 25 cm: ↑ mAs approximately 30%.

- 35 x 43 cm field size to 12 x 18 cm: ↑ mAs approximately 50%.

- 24 x 30 cm field size to 8 x 10 cm: ↑ mAs approximately 50%.

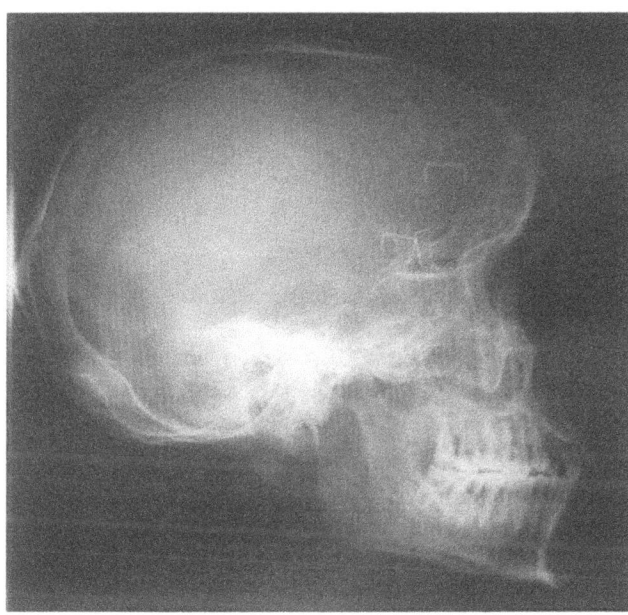

Figure 7 t: Lateral projection of skull phantom. Exposure factors: 66 kV, 24 mAs, grid, at 100 cm FFD. Adequate radiographic contrast obtained.

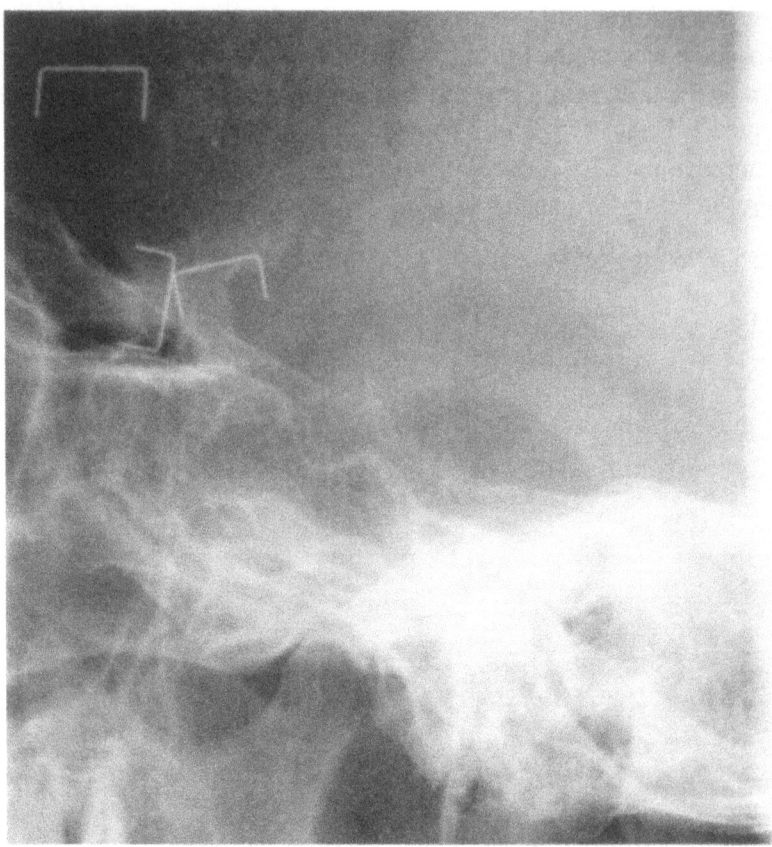

Figure 7 u: Coned lateral pituitary fossa of a skull phantom: extensive beam restriction without altering exposure factors were used for Figure 7 t. This shows that film density decreases when the film is reduced to reduce area of scattered radiation. It is necessary to increase mAs when restricting beam to compensate for loss of film density from scattered radiation.

Selection of high mAs (e.g. > 100) results in a great amount of heat production in the tube. Thus, it may be necessary to select broad focus when undertaking radiography on large patients as it may not be possible to use small focus due to tube limitations. Manufacturers include safety measures to prevent overheating of the tube although some units require one to check the tube rating and cooling charts applicable for each unit.

> **Suggested mAs adjustments when using different output generators:**
>
> **Low to higher output**
> - 2 pulse to 6 pulse, divide mAs by 1,5
> - 2 pulse to 12 pulse, divide mAs by 2
> - 6 pulse to 12 pulse, divide mAs by 1,2
>
> **High to lower output**
> - 12 pulse to 6 pulse, multiply mAs by 1,2
> - 12 pulse to 2 pulse, multiply mAs by 2
> - 6 pulse to 2 pulse, multiply mAs by 1,5

Exposure manipulation: kV/mAs

Each patient is unique in terms of size and shape and possible pathology. This requires exposure manipulation for optimal image production. There are two types of radiographic exposure charts that are used by most radiographers, namely variable kVp and fixed kVp charts, respectively.

Patient size needs to be ascertained when considering selection of kV to penetrate the anatomical structures. Measurements of a patient can be done by means of a calliper [Figure 7 v].

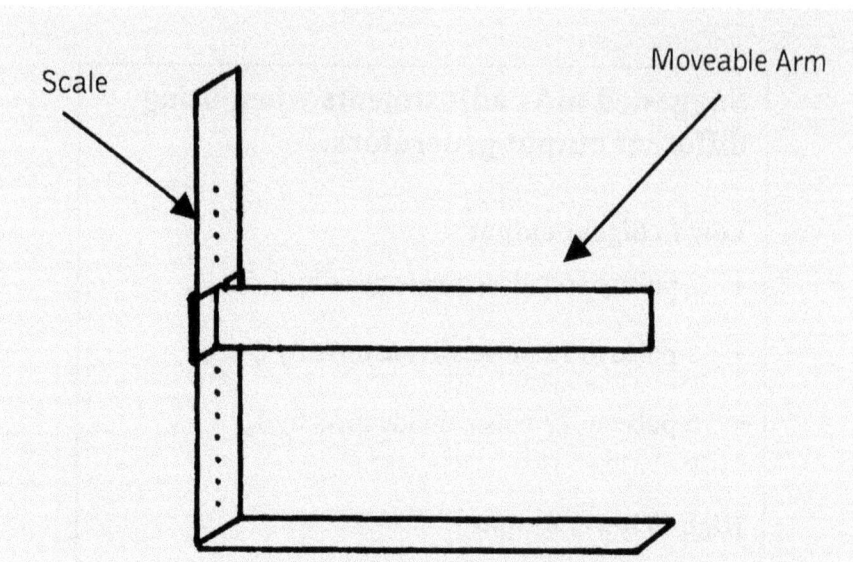

Figure 7 v: Drawing of a calliper used for calculating exposure factors based on patient's measurements as described by Lloyd in Quality assurance workbook for radiographers and radiological technologists (WHO/DIL/01.3).

Determining variable kVp charts

- Use the calliper to take measurements of patient along the line of the central ray;

- Chest measurement to be taken with patient seated or standing in a neutral position and normal respiration;

- If the part being measured is wedge-shaped, then take minimum and maximum measurements to obtain average by adding both measurements and dividing by 2.

Variable kV formula: Anatomical regions	Thickness X cm = Minimum kV
Adult bony tissue	Thickness cm X 2 + 27 = min kV
Adult chest	Thickness cm X 2 + 22 = min kV
Child bony tissue	Thickness cm X2 + 22 = min kV
Child chest	Thickness cm X 2 + 17 = min kV
Infant bony tissue	Thickness cm X 2 + 17 = min kV
Infant chest	Thickness cm X 2 + 12 = min kV
e.g. adult mid thigh measurement of 20 cm	20 cm X 2 = 40 + 27 = 67 kV

The above formula is a guide to establish minimum kV but this often means that the dose to the patient is high as the beam is not penetrating. For example, chest radiography should be done using high kV technique to reduce absorbed dose and for overall improved visualization of lung bases that are often not well seen if kV is too low [Figures 7 w and x].

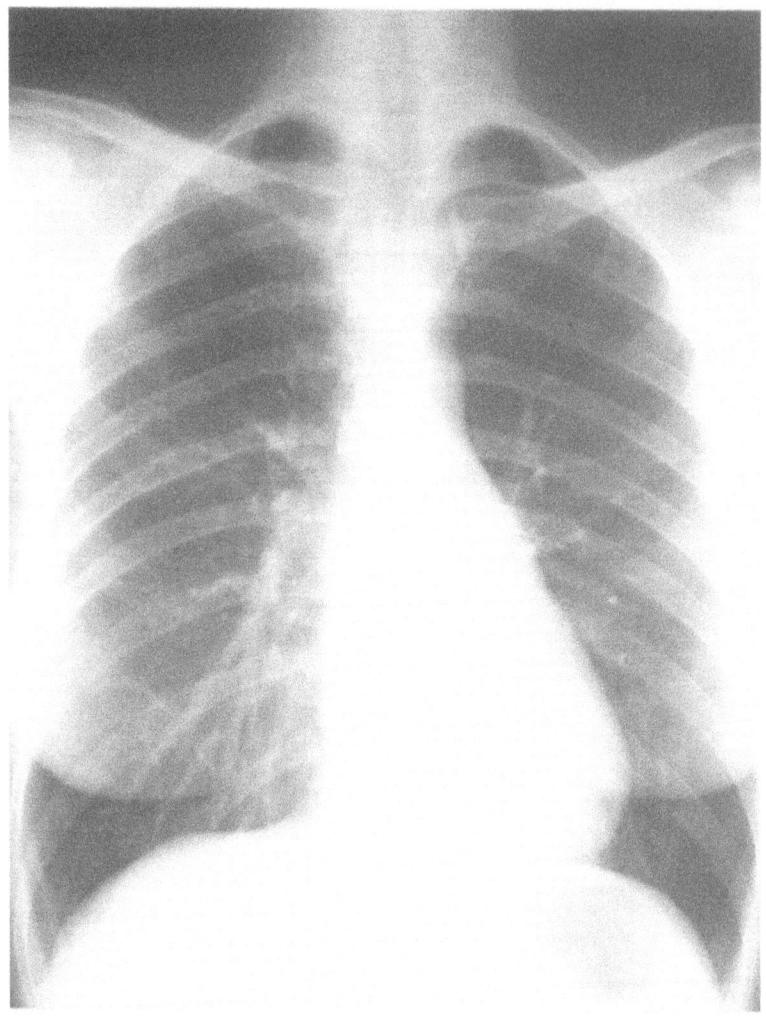

Figure 7 w: PA chest radiograph with exposure factors 70 kV, 16 mAs, 8:1 grid, at 180 cm FFD. Note: subject contrast is fairly high; lung bases not well penetrated. Compare this image with Figure 7 x of same patient.

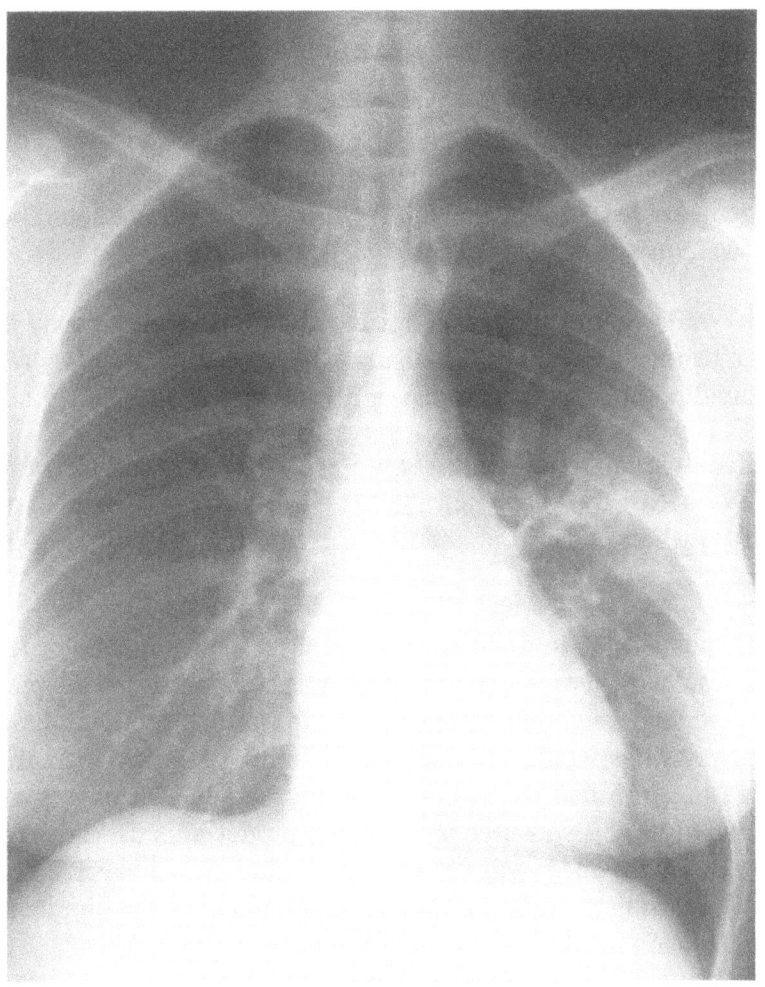

Figure 7 x: PA chest radiograph of same patient as in figure 7 w taken some years later. Note long scale contrast image obtained using 106 kV and 2 mAs. Overall improved penetration of bases with reduction of absorbed dose due to high kV and low mAs factors.

The variable kV technique requires adjustments, but it should be noted that the patient's clinical history needs to be taken into account to ensure adequate penetration. Hard to penetrate pathologies, such as sclerotic metastasis, osteoma, and radiation fibrosis require additional kV, whereas less kV is needed for more easily penetrated process, such as active osteomyelitis, degenerative arthritis, and osteoporosis

- **Fixed kV charts** are based on using a given kV then adjusting mAs to penetrate the structures. For example, adult shoulders could be examined using 65 kV, but if a patient is very obese then mAs would be increased. In this method, the dose to the patient could be high due to use of mAs for penetration of structures;

- **kV/mAs manipulation.** To use the highest possible kV may require adjustments to variable kV charts. An increase of 15% of kV requires dividing mAs by 2 to produce a comparable image;

- **Penetration of dense plaster of Paris (POP)** requires increasing the kV at least by 25% when POP is wet and freshly applied, and by 10% when POP is old and dry. Increase the mAs by 15% in both cases.

Deciding what to do in terms of exposure adjustments for repeats, often results in 'guess work' if a systematic approach is not adopted. It is therefore suggested that decisions for exposure adjustments should be based on what would be required in terms of either increasing or decreasing film density by a specific percentage. For example, if an original mAs of 4 was used and the film produced is dark, then a 25% decrease of the mAs will be 4 − 1 = 3 mAs. If a lumbar spine was examined using 80 mAs and there is a need to repeat the film due to the original being too dark, mAs should be decreased by 25% which would mean using 60 mAs. These adjustments are based on training and experience; it is recommended to always evaluate each image in terms of overall film contrast and film blackening by thinking of possible percentage changes to exposure factors.

If you are not sure what to adjust for the repeat, then the following 'rule of thumb' could be used: Increase kV by at least 15 and mAs by one third (33%) for lateral radiographs if the image quality of AP images is acceptable. For lateral projections of the skull, however, the above mentioned rule of thumb should be applied in a reversed way, i.e. the exposure factors used for the AP skull should be *decreased* for lateral projections by 15 and 33%, respectively;

- ***When in doubt,*** *increase kV instead of mAs to minimize dose to patient;*

Several factors may influence the mAs selection. Therefore, take one image and process it before taking additional radiographs. Guiding principle: keep mAs as low as possible to comply with ALARA.

Quality assurance tests to minimize unnecessary film fog

Various QA tests can be performed to minimize possible production of sub-optimal image quality due to film fog. Increased basic film fog may be due to poor safe-lighting, inadequate film storage conditions, and unacceptable processing conditions.

Safelight tests

It is important to ensure that darkroom safelight does not fog films. Unwanted film blackening (fog) causes reduction in radiographic contrast. Safelight tests should be done at least every six months to ensure they are properly working.

- Equipment for safelight tests:

An acceptable film screen light-tight cassette; black paper one-half the size of the cassette (2 sheets of black paper needed), clock (timer) with a second hand, box of unopened radiographic film, and general X-ray unit capable of selection of low mAs.

- **Step 1:** Cover lights on the processor and switch off all lights in the darkroom. In total darkness place an unexposed film in the light-tight cassette containing intensifying screens;

- **Step 2:** Expose the loaded cassette to radiation to obtain approximately the density reading of 1. Suggested exposure: half mAs of finger exposure but use same kV. Working tip: If it is not possible to select low mAs, then increase FFD using the inverse square law principle to determine the required mAs;

- **Step 3:** In total darkness open the cassette and remove the exposed film. Block off half of the exposed film using one sheet of black paper. This section of the film must remain covered throughout the test because this density is used as the control to ascertain whether the safelight is functioning correctly. Place the remaining sheet of paper on the other half of the film. Move the paper down to uncover part of the film to about midway. Switch on one safelight and expose the uncovered film portion for 60 seconds to the safelight as per normal working conditions and film handling in the darkroom. Remove this sheet of paper off the film and expose the remaining uncovered film half for further 60 seconds;

- **Step 4:** Remove the paper used to cover the other half of the film during the test. Process the film;

- **Step 5:** Place the film on an illuminator and inspect film density. This is a visual check to see whether there are differences in the densities of the film exposed only to radiation and the other half of the film exposed to radiation plus safelight. There should be very little visible evidence of increased film densities. The part of the film that was covered throughout the test, should have a density of approximately 1. Acceptable density limits when comparing the half side not exposed to the safelight and the half side exposed to the safelight, should not exceed 0.02 density for 60 seconds based on densitometer readings.

The above steps are to be repeated to check each safelight in each darkroom. Should unacceptable fogging be detected, then check (i) position of the safelight from the work bench (the height should not be less than 120 centimetres), (ii) safelight filter, and (iii) wattage of the safelight bulb.

The indicator light on the processor should be checked as per the above steps. The indicator light is to be uncovered and the safelight switched on during the tests. Density should not exceed 0.05 for a 2 minute exposure of film to indicator light plus safelight.

Processor control: performance monitoring

This quality control test should be done for all film processors to reduce unnecessary repeats. Monitoring film quality with regard to processing factors means assessing film contrast, film speed, and base fog as objectively as possible.

Equipment required

- sensitometer to expose film to different steps of light intensities;

- densitometer to measure optical density of selected sensitometric steps;

- thermometer to manually check the temperature of the chemicals;

- box of unexposed film;

- sheets of processing control charts or graph paper;

- **Step 1:** Under safelight conditions, expose one film to the sensitometer. It is important that blue light be used for monochromatic (blue-sensitive) film and green light for orthochromatic (green-sensitive) film. Process exposed film after checking temperature of chemicals as outlined in Step 2;

- **Step 2**: Temperature of chemicals to be checked using a manual thermometer. Temperature-gauge readings to be recorded. Note that both temperature readings should be the same; if there is a difference, then the faulty temperature gauge must be repaired as soon as possible to ensure the developing temperature is constant for optimal film processing;

- **Step 3**: Process the exposed film;

- **Step 4**: Using a densitometer, read densities for each sensitometric step on the film;

- **Step 5**: The sensitometric step with the density closest to 1,20 (mid-density), is to be used to determine speed index. For example, if sensitometric step 9 has density closest to 1,20 (including base fog), then all subsequent readings for speed index should be at this value (i.e. step 9);

- **Step 6:** To obtain contrast index, refer to the recorded density readings of all the steps. Since contrast is the difference between two densities, the contrast index is obtained by subtracting the density reading of one step from the readings of another step, and use is made of the density reading for the speed index, i.e. if step 9 is closest to mid-density, then subtract density reading of step 11 from step 9 to get contrast index. Same steps to be used for all future readings;

- **Step 7**: On graph paper, record temperature, date, and base fog reading;

- **Step 8**: Plot speed index and contrast index on graph paper;

- **Step 9**: For five consecutive days, repeat the preceding steps to obtain average density for the speed index and the contrast index. These become the controls against which all future sensitometric films will be compared. Plot average speed index and contrast index obtained over the five days on graph paper. An acceptable variation is **plus or minus 0.15** (on the special graph paper). Any deviation outside these lines means one or more processing factors are not performing correctly. For example, check replenishment-rates of developer, check the temperature as an increase/decrease causes changes to film density, check that chemicals have been correctly mixed in terms of specified quantities and that tarter has been added. If there is marked increase in base fog, then perform safelight test.

Careful film-handling and film storage

Exposed film is sensitive to light and care should be taken when handling film at all times.

Film should always be handled with clean hands and in a dust free environment. Film boxes must be placed vertically (i.e., "standing") in a cool room with good air circulation.

Boxes of film must never to be stacked flat on top of each other as this will cause marks on the films. Pressure, even very slight, on film causes 'white streaky' marks that could be incorrectly interpreted as pathology.

Summary

To reduce unnecessary radiation dose to patients, optimal image quality should be obtained using the highest possible kVp for visualization of the anatomical parts to be examined. If the examination requires visualization of soft tissues, low kV technique should be used. Patient dose is decreased when overall penetrating abilities of the X-ray beam are high (high kV). Patient dose is increased when excessive mAs selections are used. A radiograph which is very black, indicates that the patient received unnecessary radiation due to high mAs factors.

To minimize the need for repeats, it is recommended that basic quality assurance tests be undertaken. Film fog contributes to poor image formation and may result in non-visualization of detail. Illuminators (viewing boxes) should be clean and, if possible, each of them should have the same amount and type of illumination.

CHAPTER 8

Exposure to ionizing radiation during pregnancy

Patients and medical staff

Possible risks to the embryo and foetus must be considered when using ionizing radiation to investigate female patients of reproductive age. Alternative imaging modalities and techniques not involving ionizing radiation should also be considered.

A female patient of reproductive age who presents for an examination in which the pelvic area will be irradiated should be asked whether she is, or might be pregnant. If she is not sure, then it is advisable, except in an emergency, to reschedule the examination until a pregnancy has been excluded.

Pregnant radiation workers

Occupational exposure of pregnant radiation workers (radiographers/technologists) must be as low as possible, and complying with national laws and international recommendations. Most employers accommodate pregnant radiographers by allowing them to work in low risk workstations. The ICRP recommendations state that pregnant diagnostic radiographers should not be involved in examinations using mobile X-ray units, radiographic procedures in operating theatres, and fluoroscopic procedures. Pregnant radiographers should be provided with radiation monitoring devices in addition to the personnel monitoring devices, such as TLDs (thermo luminescent dosimetry).

Summary

Whenever possible, X-ray examinations of a pregnant woman should be postponed until after delivery. If the examination has to be carried out immediately, the radiation dose must be kept at a minimum, without comprising treatment of the patient. Further information can be found in the joint WHO/ICRP publication (WHO/DIL/02.1): "Ionizing Radiation for Diagnostic Imaging or Treatment during Pregnancy: Some practical advice".

Pregnant radiographers should be provided with additional radiation monitoring devices, and restricted to work in low risk areas as per the recommendations of the ICRP.

CHAPTER 9

Self-evaluation of images: application of ALARA

Radiation protection to reduce dose to staff, patients, and members of the public is achieved by legislation and education. The responsibility of implementing legislation resides with the national authorities. However, operators of ionizing radiation units are responsible for keep their work in line with the ALARA principles.

It has been stated that "good radiological practice is something that can be taught". This point of view is re-iterated by Rehani, namely that nearly 40% dose reduction is achievable by appropriate training. Hopefully this book provides some additional knowledge on this. This chapter presents a series of images to be used for assessing image quality, and to see how the ALARA principles can be applied.

Practical hints for self-evaluation of image quality and implementation of radiation protection measurements

Carefully assess Figures 9a to 9i and read the accompanying text for each figure to test your knowledge of factors affecting good radiographic practice, such as FFD, beam restrictions, and gonad protection. Turn to the end of the chapter for suggested answers.

- Figure 9a: Lateral thoracic spine.

- Figure 9 b: Lateral skull.

- Figure 9 c: Lateral skull.

- Figure 9d: AP abdomen.

- Figure 9 e: AP abdomen.

- Figure 9 f: AE (after evacuation) film.

- Figure 9 g: Delayed AE film.

- Figure 9 h: Lateral abdomen.

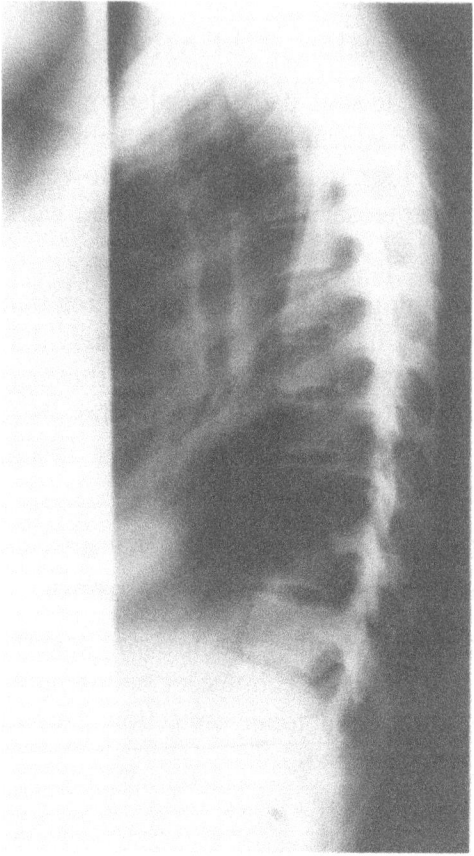

Figure 9a (left): Lateral thoracic spine. Is this a good image? Is there evidence of implementation of ALARA? Explain why the ribs are blurred. Why is there a lack of penetration and visualization of the upper lumbar vertebrae?.

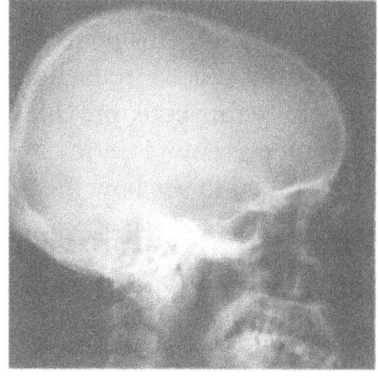

Figure 9 b (above): Comment on this lateral skull in terms of contrast and image detail. Would you repeat this radiograph?

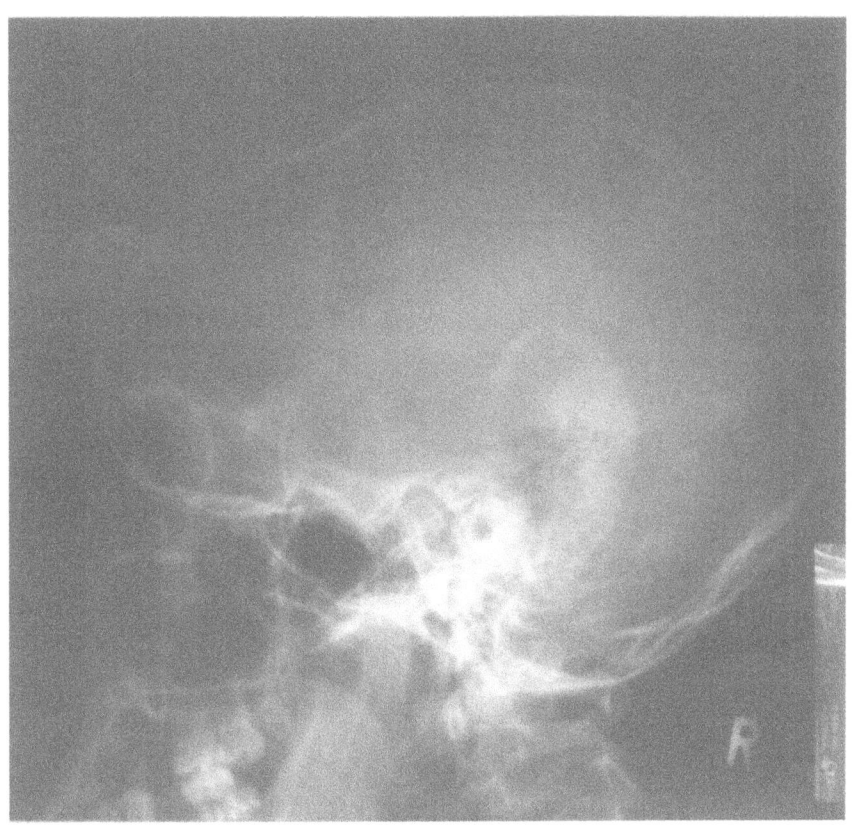

figure 9 c: Would you repeat this lateral skull radiograph? Consider ALARA and benefits to patient.

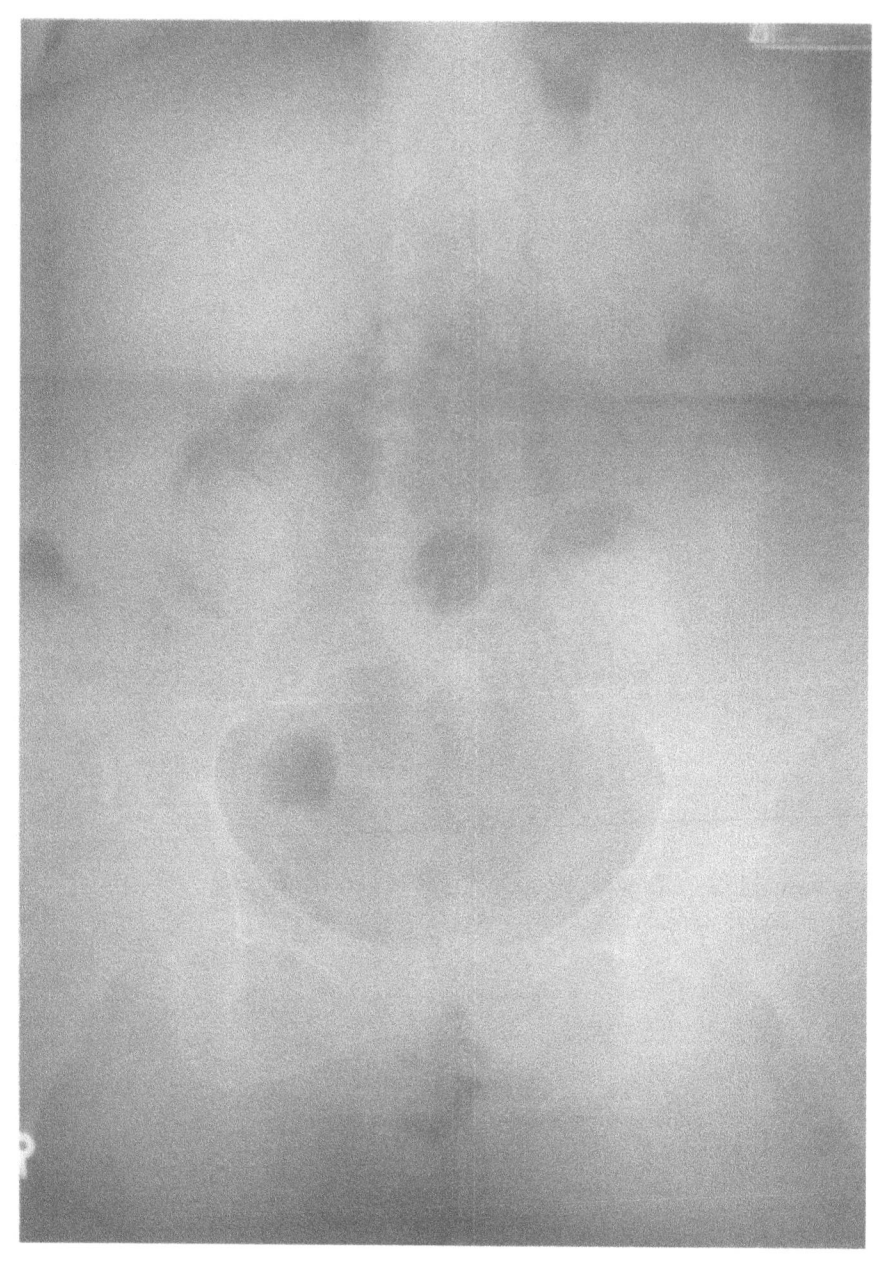

Figure 9d: AP abdomen. Which factors would you alter to achieve improved penetration of the abdomen? Would your suggested factors increase or decrease patient's absorbed dose?

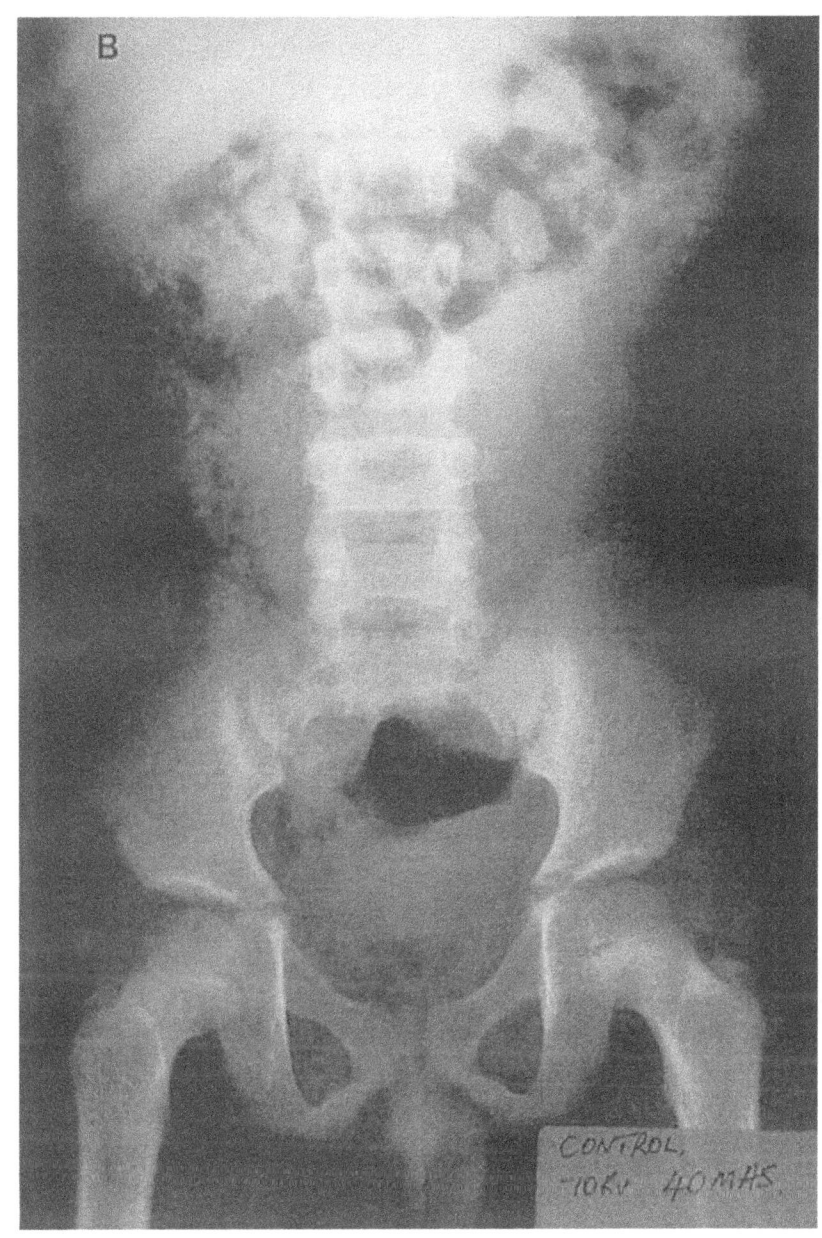

Figure 9 e: AP control abdomen of 10 year old male for barium enema. Is this image in keeping with ALARA?

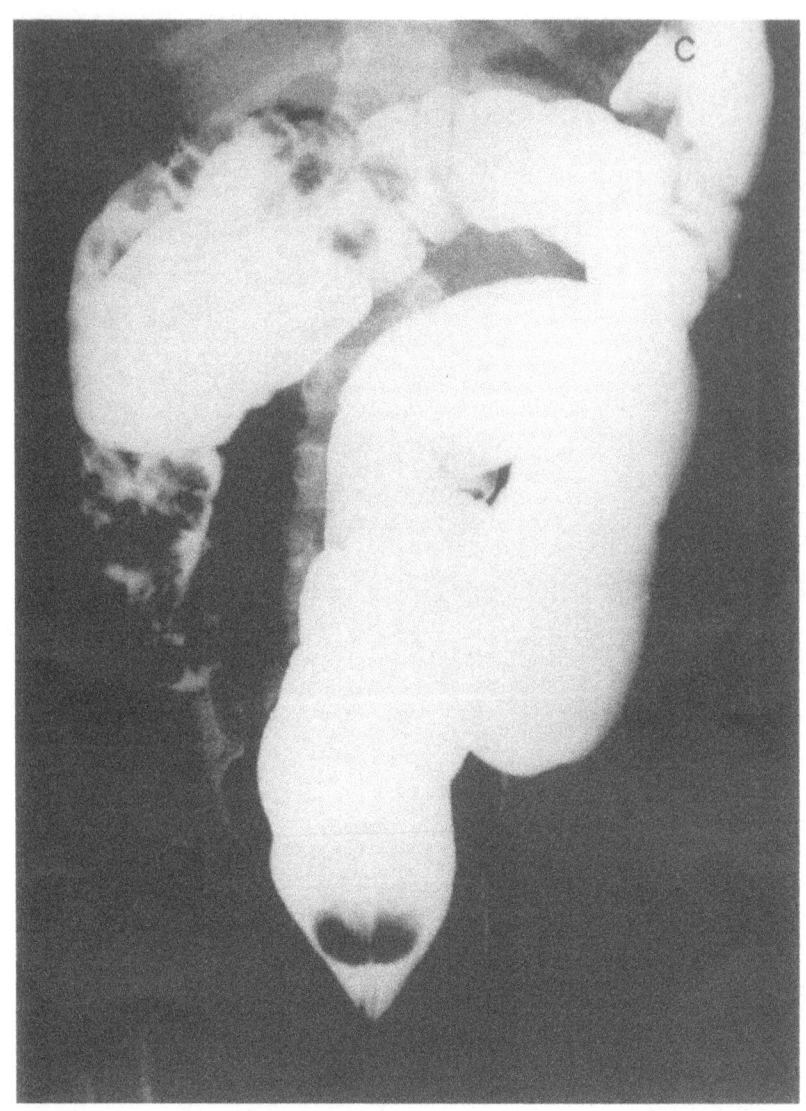

Figure 9 f: AE (after evacuation) film of same patient. Is this radiograph acceptable based on the ALARA principle?

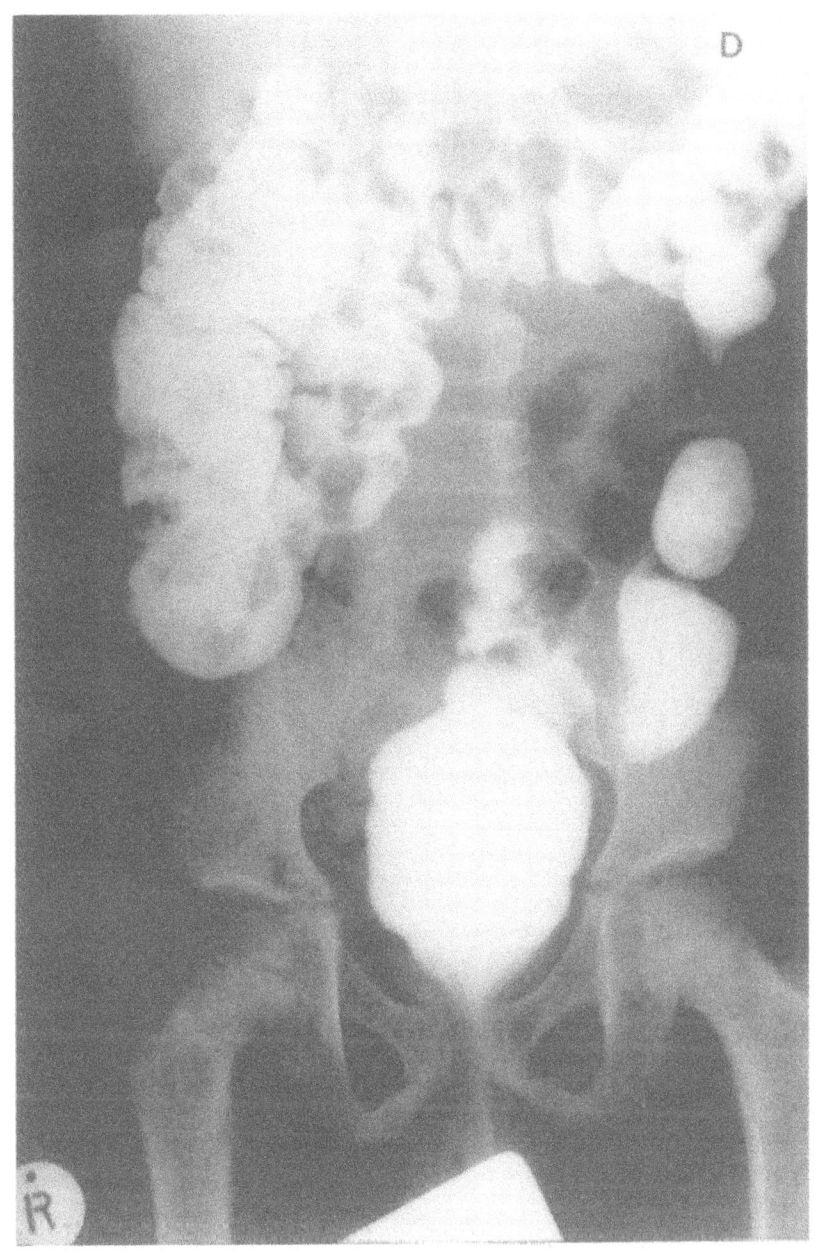

Figure 9 g: Delayed AE film of same patient. Note evidence of gonad protection on this long scale contrast image. Is this image in keeping with ALARA?

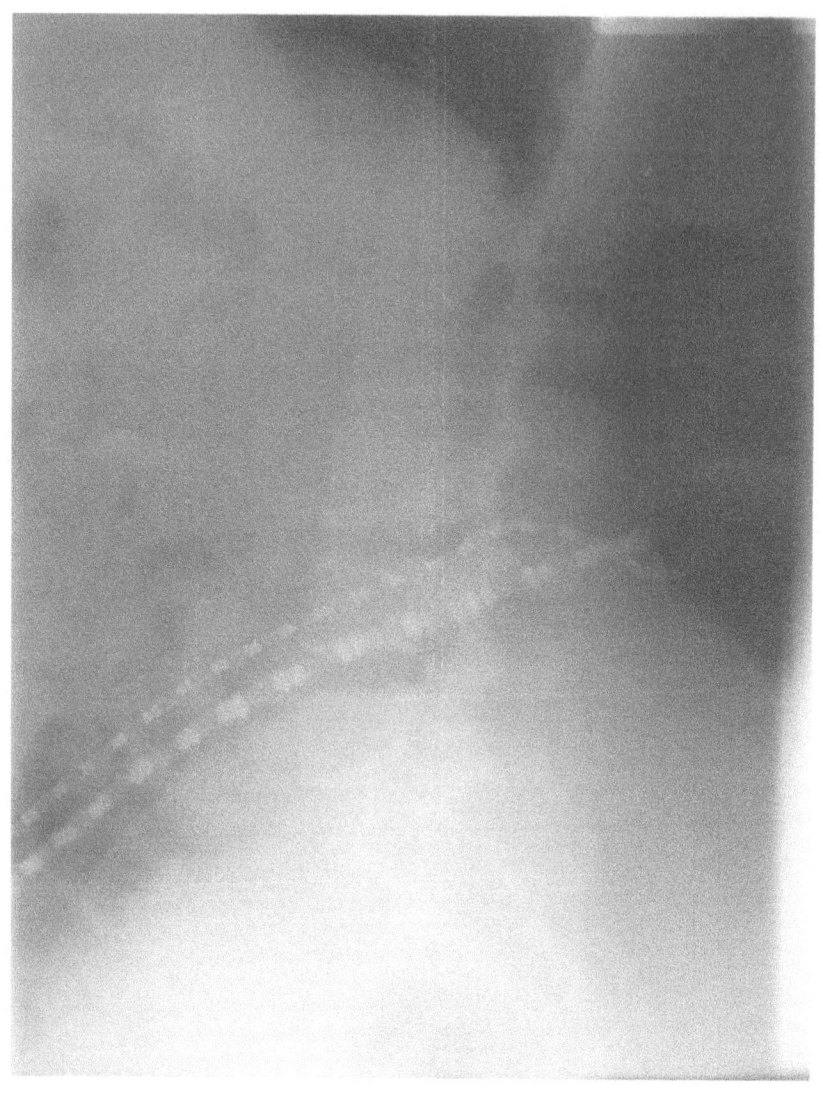

Figure 9 h: Note the detail of some beads is blurred. Why? Do you position a patient in a specific way to produce blurring of some anatomical structures in keeping with ALARA for good image detail and /or dose reduction?

Suggested answers to questions for Figures 9a to 9h.

- Figure 9a: Image acceptable. ALARA evident as beam restricted to area of interest. Ribs blurred diffusion technique (long exposure time with patient breathing) used to blur out ribs that obscure vertebrae. Lumbar vertebrae not penetrated because they are below the air-filled lungs.

- Figure 9b: Radiographic contrast and detail good. Not necessary to repeat the radiograph even though there is no evidence of beam restriction.

- Figure 9c: This radiograph should be repeated as patient was rotated during the exposure and overall film density is increased due to high mAs. Beam needs to be collimated.

- Figure 9d: Short scale image as not sufficient kV used. Radiograph is under-penetrated meaning that kV should be increased by at least 20% for optimal visualization of abdominal structures. Increasing the kV will result in a more penetrating beam with less radiation dose to the patient. Compression could be consisectiondered to reduce overall thickness of tissue thus reducing scattered radiation.

- Figure 9e: ALARA not applied as gonads of the young male not protected. Contrast short scale due to low kV and high mAs exposure factors which meaning dose to patient not limited.

- Figure 9 f: Gonads not protected and high kV/low mAs factors to reduce dose not used.

- Figure 9 g: Application of ALARA principles is evident; gonads protected and high kV technique used as long scale contrast image with overall improved visualization of bowel patterns.

- Figure 9 h: String of beads around patient's waist not removed; in the lateral position the beads closest to the film/tabletop well seen. The beads closest to tube, due to increased subject film distance, are blurred. Examples of using this principle of geometric un-sharpness include PA mandibles, and PA sterno-clavicular joints.

www.ingramcontent.com/pod-product-compliance
Ingram Content Group UK Ltd.
Pitfield, Milton Keynes, MK11 3LW, UK
UKHW051523180426
11947UKWH00018B/1551

Hypnotherapy Training

*An Investigation Into The Development
Of Clinical Hypnosis Training Post-1971*

by

Shaun Brookhouse, **PhD, DCH,
MA, MSc, PGDHP, MHyp, MHP, FHRS**
*Recipient of the Hypnotherapy Research Society's 1998 Special Award For
Contribution To The Profession Of Hypnotherapy*

Foreword by Sue Washington, MSc, BA, Cert Ed, PGDHP, MHP, FCRAH, FCAP
Principal, Centre Training School of Hypnotherapy and Psychotherapy

Introduction by William Broom, FNCP, MNCH (Acc)
Secretary, National Council for Hypnotherapy

Crown House Publishing
www.crownhouse.co.uk

First published in the UK by

Crown House Publishing Limited
Crown Buildings
Bancyfelin
Carmarthen
Wales

© Shaun Brookhouse 1999

The right of Shaun Brookhouse to be identified as the author of this work has been asserted by him in accordance with the Copyright, Designs and Patents Act 1988.

First published 1999. Reprinted 2004.

All rights reserved. Except as permitted under current legislation no part of this work may be photocopied, stored in a retrieval system, published, performed in public, adapted, broadcast, transmitted, recorded or reproduced in any form or by any means, without the prior permission of the copyright owner. Enquiries should be addressed to Crown House Publishing Limited.

British Library of Cataloguing-in-Publication Data
A catalogue entry for this book is available from the British Library.

ISBN 1899836179

LCCN 2002116894

Printed on demand and bound by
Lightning Source UK Ltd.
Milton Keynes

For

Joseph K. Conlon

and

Silas

About the author

Dr. Shaun Brookhouse holds advanced degrees in Psychotherapeutic Counselling, Clinical Hypnotherapy, Psychology, Education Studies and Behavioural Science. He is the Director of Brookhouse Hypnotherapy, a private hypnotherapy practice, Co-Founder and Director of Training and Research of the Washington School of Clinical and Advanced Hypnosis, and Registrar of the Centre Training School of Hypnotherapy and Psychotherapy. He holds several certifications and diplomas in hypnotherapy, psychotherapy, NLP, and Time Line Therapy™, from both the UK and the USA. He is a Certified Instructor of Hypnotherapy and a Certified Trainer of NLP.

Dr. Brookhouse is founding Chairman of the UK Confederation of Hypnotherapy Organisations and is on the Advisory Board of the Time Line Therapy™ Association in Hawaii, USA and is an Editorial Board Member of the *Journal for The Hypnotherapy Research Society*. For his contributions to the field of hypnotherapy he has received 5 Fellowships and several special awards including the ***Hypnotherapy Research Society's 1998 Special Award for Contribution to the Profession of Hypnotherapy***.

Shaun runs seminars on various aspects of hypnotherapy and NLP internationally.

Table of Contents

Acknowledgements iii

Foreword v

Introduction vii

Abstract ix

Chapter 1:   Hypnotherapy 1

Chapter 2:   Background Issues 3

Chapter 3:   Context Of Study 11

Chapter 4:   Methodology 13

Chapter 5:   Findings: Data/Analysis 17

Chapter 6:   Discussion 25

Chapter 7:   The American Experience 33

Chapter 8:   The UK Federation Of Hypnotherapy Organisations 37

Epilogue 45

Appendix 1: The Foster Report 47

Appendix 2: 57

Appendix 3: 59

Appendix 4: 61

Appendix 5: 63

Bibliography: 67

Acknowledgements

I would like to thank my family and friends without whose support I would be nowhere. Also, I would like to thank my students and clients, from whom I learn more and more every day. I would like to thank Sue Washington and Bill Broom for their friendship and professionalism. Finally, I would like to thank all of the schools, colleges, and associations who helped me with my research and answered many of the questions that I had no answers to.

Foreword

In July 1968 I left my lovely college in Yorkshire, now Leeds University College, armed with a Certificate in Education and, as my now late father had wished, a job for life in teaching. Little did I know of the drastic changes that summer would bring.

A hypnotherapist had moved into the locality in private practice. It was unheard of in those days. I remember sitting in the back garden giggling with my mother and her friend whilst we joked about who should go and see him to suss him out. The subject of him stayed in my head for the next week or two, and the pull of the subject and its power to help was such an enormous draw that I decided to go and see him.

He charged me £2 for a session. My new job gave me £800 per annum, and I decided to see him again. I was fascinated in every sense. I remember saying that I was fascinated. In the best traditions of metaphor, he told me that one young man had already said that to him and that his reply was, "If you really are, you will stick to me like glue". I decided to do the same. That is thirty years ago almost to the day that I write.

He became my friend, teacher, mentor, and partner for the next nine years. I was twenty-one and he was forty-two.

I remember seeing *Brideshead Revisited* on the television years later. Our hero, Charles Ryder, talks about going up to Oxford where he meets Sebastian. He talks about "the low door in the wall to which there was no key". It led through into a magic place in which all sorts of exciting things happen. He has heard of it but has no access. He doesn't know the way in – until meeting with Sebastian.

That was the place Peter Blythe took me to. He was on an international scene, and I, at my age, should barely have been on a local one. For instance, I got to know the late Dr John Hartland, MD, the author of that still most widely accepted standard textbook of our subject **Hartland's Waxman's Medical And Dental Hypnosis**, and I remember him with great affection. One of the most notable memories is of a journey on a Swedish train at the end of a fortnight's University lecturing tour for the three of us. The guard had only one compartment left, with three sleeping bunks stacked one over the other. John Hartland "Couldn't sleep on trains".

Deciding that drastic action needed to be taken to help out this adorable old man, I poured him a sympathetic gin and something – which we all shared out of a tooth mug. I slept like a filling in a sandwich, with Peter Blythe over the top of me and John Hartland below. John awoke not believing how well he had slept.

Peter Blythe already trained doctors, and shortly afterwards started a training group for both physicians and lay people that lasted three or four years. We were insatiable. There were people in that group who are still around and well known today, such as Geoff Graham, a past president of the British Society of Medical and Dental Hypnosis. After we had finished our course, Peter formalised that training into a year's course, and started the Blythe Tutorial College of Hypnosis and Psychotherapy with a partner, Harry Bailey-Marsden. He then sold it to a consortium, and a member of this is still the owner today.

In 1983 I formed Centre Training School of Hypnotherapy and Psychotherapy. Our training extends as far as South Africa, and we are negotiating in Tanzania, Greece, and Dublin. Taking a model of this training to help the world become a better place, I made a model to teach and share – forming the companies of Peace of Mind Training Ltd and Peace of Mind International Ltd on the way. Life is exciting.

Several years ago, I met Dr Shaun Brookhouse. I remember our first conversation as if it were yesterday. There were many more conversations before Shaun was eventually persuaded of the importance of UKCP Registration. He subsequently decided to retrain with Centre Training School, to gain UKCP validation. He has since become a good friend and business partner. He is one of the most hard-working, reliable, and "solid" people I have ever met. I remember once needing some information about something or other. He wrote back to me with the details, affectionately signing, "Ever your ferret, Shaun".

We speak often on the telephone now. Recently I referred to him as the "Political Wing" of hypnotherapy. No one in the country has such an admirable grasp as he on the politics of the subjects of hypnosis and hypnotherapy. Shaun certainly "ferreted" and fulfilled his task admirably to bring you the following volume. I recommend it to you wholeheartedly and without reserve.

Sue Washington, MSc, BA, Cert Ed, PGDHP, MHP, FCRAH, FCAP
Principal, Centre Training School of Hypnotherapy and Psychotherapy
May 1998

Introduction

One of the main obstacles to the emergence of Hypnotherapy as a unified, discrete profession in its own right has been the plethora of available training courses and the concomitant absence (ie. until recently) of any agreed standards on training criteria.

This has effectively resulted in the reluctance of General Practitioners and other relevant third parties responsible for patient care, to refer suitable patients to hypnotherapists, due to an inability to determine which practitioners may be adequately "qualified".

Shaun Brookhouse's investigation, and subsequent report on his findings, is therefore an essential preliminary to the work which our fledgling profession must undertake if it wishes to mature into the fully developed "respectable" occupation so desired by practitioners and deserved by the public who seek our services.

Shaun's has been an unenviable task, given the historical suspicion that exists among training course providers for anyone attempting to research their essential structure and ethos, yet a task which has been carried out with his customary fortitude and tenacity.

The report's findings are, in many respects, encouraging, not least by demonstrating the commitment displayed by some of the main training providers both to offer a thorough tuition to students and to voluntarily submit to external validation by one or more respected professional bodies.

This is a decisive move, by responsible course providers, away from the rather narrow and insular philosophy that has previously prevailed and one that can be further advanced by this timely and authoritative report by a practitioner, trainer, and now researcher, who justifiably enjoys a well-earned reputation as one of the most knowledgeable members of this most complex profession.

William Broom, FNCP, MNCH (Acc)
Secretary, National Council for Hypnotherapy

Abstract

Since 1954, the British Medical Association has recognised hypnosis as a valuable therapeutic modality (Brookhouse, 1997). However, in the past 26 years it has grown in popularity like never before. There are many training colleges in hypnotherapy, but most of them are run in the private sector and have not come under any governmental or academic scrutiny. The two degree courses in applied hypnosis are run exclusively for the medical, dental and psychological professions.

It has become clear that there needs to be some sort of understanding between the medical, dental and psychological professions and those who are called "lay" hypnotherapists. This expression refers to those who are in practice with no medical or psychological training. It is my belief that the way to achieve this understanding is through an agreement on training.

At present, there is a variety of initiatives relating to accreditation of training courses. These range from self-accreditation to accreditation by independent professional associations and registers, to the development of occupational standards (NVQ's). The purpose of this paper is to set out the variety of arguments regarding rights of practice, and the historical antagonism between the medical and psychological professions and the "lay" hypnotherapist, and finally to determine whether there is some common ground in the training of hypnotherapists which will allow some sort of co-existence to flourish.

Chapter 1

Hypnotherapy

Before one can look at the aspects of training in clinical hypnosis, one must first understand what hypnosis is.

"Clinical hypnosis is a skill of using words and gestures in particular ways to achieve specific outcomes." (Yapko, 1990, p.4)

Part of the difficulty hypnosis has had is due to the misconceptions held by many members of the general public. These misconceptions about being "put under" or "being unconscious" are mainly attributable to the movies and to stage hypnotists. Also, as hypnosis has to do primarily with the mind, opponents to lay practitioners make a convincing argument that only those who are fully accountable for the clinical care of a patient should be allowed to practise hypnosis.

However, in my view, as both a practitioner and a researcher, hypnosis is no more dangerous than a daydream. It is a marvellous state of mind where anything is possible. In fact, hypnosis is actually induced by the subject and not the hypnotist (Alman and Lambrou, 1992). If this is the case, why then should we be worried about the training of hypnotherapists at all?

Unlike in many other Western countries, the practice of hypnosis is largely unregulated in the UK. There have been only three occasions where hypnotism has attracted Parliament's attention: The 1952 Hypnotism Act, which regulated Stage Hypnosis. In 1971, the government commissioned Sir John Foster, KBE, QC, MP, to compile a report into the practices of Scientology in relation to psychotherapy and hypnosis. Finally in 1980, there was a failed attempt to strengthen the 1952 Hypnotism Act.

It is the Foster Report (see Appendix 1, 260) that I find to be particularly interesting. Among his many recommendations was the proposal that psychotherapy, including hypnotherapy, should be under legislative control.

(a) *"Clearly, it is only the practice of psychotherapy for fee or reward in cash or kind, paid by or on behalf of the patient, which needs to be controlled, since in a very wide sense we all practise some kind of psychotherapy on each other in our personal relationships, and many voluntary organisations try to help people with 'counselling'."* (Foster, 1971, p.180)

It seems ironic that most of the schools providing training in hypnotherapy were founded well after the publication of this report. Following Foster, two predominant views emerged regarding the training and practice of hypnotherapy. One view was that only medical, dental, or psychological practitioners should be eligible to train or practise. The other point of view was that anyone should be allowed to train or practise, and these are often referred to as "lay" hypnotherapists, who view themselves as bona fide professionals.

There are somewhere in the region of 70 organisations that profess to be the representative body for clinical hypnosis. The Royal Society of Medicine has its own division, The British Society of Medical and Dental Hypnosis, to monitor hypnosis practitioners who are medically qualified. The psychological community is represented by The British Society of Experimental and Clinical Hypnosis.

The lay community is represented by the remaining bodies. However, the three organisations with the highest profile are: The National Council for Hypnotherapy; The British Register for Complementary Practitioners Hypnotherapy Division; and The UK Council for Psychotherapy Hypno-Psychotherapy Section.

There are six potential problems with acceptance by the medical, dental and psychological communities of lay-trained hypnotherapists. They are:

- Lack of Common Course of Study
- The Variance of Entry Requirements
- The Variance of Length of Courses
- The Variance of Awards Given
- The Stated Goal of the BSMDH and BSECH to Ban Lay Hypnosis Practitioners
- Lack of University Validation

The purpose of this report is to investigate whether or not there is a chance for "lay" hypnotherapy training to become more widely recognised in the medical, psychological and educational communities.

Chapter 2

Background Issues

To gain a better understanding of the issues surrounding the training in Clinical Hypnosis, I have taken the past 5 decades and catalogued the major issues. The main headings are: Growth of Associations; Training; Quality Assurance; and Politics and Culture.

The 1950s and 1960s

Growth of Associations: In 1953, the Psychological Medicine Group Committee of the British Medical Association looked at the viability of hypnosis as a therapeutic modality (Waxman, 1989). The report, published in 1955, concluded that:

"A description of hypnosis and of its psychotherapeutic possibilities, limitations and dangers, be given to medical undergraduates, and instruction in its clinical use be given to all postgraduate psychiatric trainees and possibly to trainee anaesthetists and obstetricians." (Br. Med. J. (Supplement), 1955, p.190)

The above recommendation became a recognition by the British Medical Association of the validity of hypnosis as a therapeutic discipline.

In 1952, the British Society of Dental Hypnosis was founded. This was to become, in 1955, the Dental and Medical Society for the Study of Hypnosis. Finally in 1961, after an amalgamation, the Society for Medical and Dental Hypnosis was formed.

The Society changed its name in 1968 to the British Society of Medical and Dental Hypnosis (Waxman, 1989). The BSMDH still exists in this form today. There is even a Section of Medical and Dental Hypnosis in the Royal Society of Medicine.

In the 1950s and 60s this was the only hypnosis association of any kind in the UK. Many psychologists who practised hypnotherapy joined the International Society for Experimental and Clinical Hypnosis in the United States. There was no organisation in existence for the lay practitioner of hypnosis. This is largely due to the fact that there were very few lay practitioners around at this time.

Training: Even though the BMA recognised the validity of hypnosis, the subject was largely ignored by Medical Schools and Universities (Heap and Dryden, 1991). So the private sector became the provider of hypnosis training. This leads to a very interesting state of affairs. With so few physicians and dentists being able to train others in hypnosis, the private, lay hypnotherapy school was born. In the late 1960s Peter Blythe, a lay hypnotist, founded the Blythe College of Hypnosis and Psychotherapy. This school, in a new incarnation, The National College of Hypnosis and Psychotherapy, still exists today.

Quality Assurance: In the 1950s and 60s quality assurance did not exist. However, it was ensured, in so much as the fact that only physicians and dentists tended to practise hypnosis. Therefore, by implication, the medical and dental professions looked after the practice of hypnosis through parliamentary statute.

For the few non-medically trained hypnotherapists there was no validation of training or monitoring of practice at all.

Politics and Culture: As far as the 1950s and 60s went there was only one interest in hypnosis by the general public. That was the practice of hypnosis for entertainment purposes – stage hypnotism. In 1952, this came under parliamentary scrutiny in the form of a court case, Rains-Bath v Slater. (Waxman, 1989)

Ralph Slater was an American hypnotist who performed in Brighton in 1948. During this performance a lady accused Slater of assault and professional negligence. He was found guilty of professional negligence but not of assault (Singleton, Lord Justice 1952). This incident led to a private member's Bill being passed in Parliament. In August 1952, the Hypnotism Act was placed on the statute book. The Act conferred power to any local authority which granted licences for the regulation of places used for public entertainment, to attach conditions to that licence in relation to the demonstration or performance of hypnosis (HMSO 1952).

The 1970s

Growth of Associations: From the late 1960s there was a growing interest in hypnosis by both the lay and psychological communities. Until now, there was no professional organisation for the lay practitioner in the UK. The publication of the Foster Report in 1971 changed that.

Though the Report was initiated by alleged abuse in the Scientology Sect, it became apparent that the practice of psychotherapeutic techniques by those not qualified in medicine or psychology would also be reviewed.

The main hypnotherapy school of the day had a professional association linked directly to it, but in the early 1970s there was no independent body for non-medically and non-psychologically qualified practitioners of hypnosis.

In reply to the recommendations of Foster, a letter was written to *The Daily Telegraph* stating:

"I would welcome the establishment of an Association of Ethical Psychotherapists with whom the Government could deal and which could set and apply standards of practice, thus increasing the benefits of the service we render to the community. This letter is, therefore, to ask all interested hypnotherapists whose livelihood is endangered by the Report to contact me by phone or letter. I would indeed be grateful if people who have been aided by hypnotherapy (or indeed feel they have a complaint) contact me." (As cited in Cousins, 1995, p.3)

This letter became the start of the National Council of Psychotherapists and Hypnotherapy Register. This eventually split and became the National Council of Psychotherapists and the National Council for Hypnotherapy. This organisation is still in existence today.

A few years later, in 1978, the British Society of Experimental and Clinical Hypnosis was founded (Heap and Dryden, 1991). This organisation was established to represent psychologists, primarily, and physicians and dentists in the experimental and clinical uses of hypnosis. This organisation is still in existence today.

Training: In the 1970s there was a significant increase in the number of hypnosis training providers. These schools, like Blythe College before them, trained not only recognised health professionals, but also lay people with various degrees of skills and abilities. One could say that this was a good thing, bringing fresh ideas into a fledgling field. Others say that the lack of public accountability of these courses and those who graduate from them means:

"...the public has little or no protection against a variety of potential dangers." (Heap and Dryden, 1991, p.197)

With the advent of the British Society of Experimental and Clinical Hypnosis (BSECH), there has been a greater attempt to get universities involved with training professionals in hypnosis.

Quality Assurance: As in the 1950s and 60s there was no quality assurance, as such. However, with the advent of bodies like the NCP and HR and BSCEH, there was more self-regulation beginning to emerge. By "self-regulation" it is meant that these organisations had an embryonic system relating to code of practice, minimum training requirements, and complaints and disciplinary procedures.

As the field grew in the 1970s there were increasing calls for some form of statutory regulation (Waxman, 1989).

Politics and Culture: The general public were still fairly ignorant of the applications of hypnosis. Stage hypnosis still had a high profile in this period, but, because of the techniques of indoctrination employed by Scientologists, the public became concerned about the practice of psychotherapeutic techniques by those unaccountable to the law (Heap & Dryden, 1991).

The 1980s

Growth of Associations: The 1980s saw an explosion in the popularity not only of hypnotherapy, but also of the other so called Complementary Therapies (Boye-Thompson, 1996). Several organisations were founded in the 1980s. Many did not last more than a few years while others, like the Association for Professional Therapists (APT) and the Institute for Complementary Medicine (ICM), still exist today. This explosion was due to the growing demand for complementary therapies and the losing of faith in conventional forms of treatment (Boye-Thompson, 1996). Many of the hypnotherapy associations that were founded in the 1980s were directly tied to a training organisation. This use of "associated professional bodies" helped to add credibility to courses that could not or would not get validation from other sources. (Brookhouse, 1995)

As far as the medical and psychological societies of hypnosis went, there was very little happening. The exception was the beginning of a campaign to legislate against lay practitioners (Heap and Dryden, 1991). In fact, there appeared some published statements that lay hypnotherapists were not qualified to practise (Waxman, 1989).

Training: The growth of training courses went hand-in-hand with the growth of associations. From a relatively small number in 1980, the field of course providers grew to over seventy by 1989. Also, there was an increase in the 1980s of correspondence courses in hypnotherapy. The multitude of possible courses made many in the BSMDH and the BSECH uneasy.

In fact, it led to some organisations making claims that appeared to be false in various hypnotherapy publications.

"First of all it must be understood that there is no authentic degree or diploma in hypnosis issued in either Great Britain nor in any other country in the world. There is no such thing as a 'qualified' hypnotherapist..." (Waxman, 1989, p. 480)

Because of this, now, open hostility between various sections of the medical and psychological communities with the lay hypnosis community, the need for a recognised course structure began to be recognised in the late 1980s.

Quality Assurance: As in the previous three decades, there was no formal quality assurance for hypnotherapy courses in the 1980s.

However, the need for it, especially in the lay community, began to become desirable. In the mid 1980s the ICM began to accredit training courses in complementary therapies. The first division to be established by the ICM was the hypnotherapy division. Four courses achieved accredited status with the ICM by the late 1980s (Boye-Thompson, 1996). Lay associations began to develop complex codes of practice and accreditation procedures (Appendix 5). Complaints procedures at this stage became less idiosyncratic with written procedures and set time scales (Appendix 4). Despite these attempts to self-regulate the field, there was still open hostility from the BSMDH and the BSECH.

Politics and Culture: The 1980s began with an attempt to strengthen the 1952 Hypnotism Act. Though the Act primarily dealt with stage hypnosis, the proposed strengthening of the Act would have practically eliminated the scope lay practitioners currently have to practise (Kinnoull, 1979). The amendments to the Hypnotism Act had the support of the BMA, BDA, The Law Society and a variety of Royal Colleges. However, the Act did not get past a second reading in the House (Waxman, 1989).

"One need only consult the classified telephone directories of cities throughout the country to realise that the medical profession is assisting this new crop of pseudo medical victimisers of the general public by all too often shutting its eyes to hypnosis..." (Erickson, 1980, p.537)

The lay hypnotherapy profession was far too fragmented at this time to mount any kind of campaign to lobby Parliament for favourable legislation. In fact, at this stage the field was not even sure if legislation was the course to follow.

The 1990s

The Growth of Associations: From the explosion of new associations in the 1980s, the situation seems to have stabilised in the 1990s. Though the hostility still exists in certain sections of the medical and psychological communities, the 1990s started with a more conciliatory tone by the British Medical Association. The BMA addressed several non-conventional therapies and approached the main bodies representing them (BMA, 1993). There seemed to be some common ground among the main players in the debate on the importance of associations.

"There appears to be a general drift among therapists, parliamentarians, and others away from umbrella bodies toward single registering bodies for each therapy." (BMA, 1993, p.70)

It was during the 1990s that the NCH grew from 210 members to 847. Through the main independent associations, there seemed to be a recognition that some form of accreditation to ensure the public good was now necessary. Also, the United Kingdom Council for Psychotherapy launched its first National Register of Psychotherapists in 1993.

Hypnotherapy has its own Section within the Council, which has since been renamed The Hypno-Psychotherapy Section. Though this is a relatively small Section within UKCP, it does allow hynotherapists to look towards being trained to European Standards through UKCP's membership of the European Association for Psychotherapy.

The BMA Report led to a few skirmishes between associations to get the favour of hypnotherapists who were undecided as to whom to align themselves with. At present there are only two major independent hypnotherapy associations in the UK, the National Council for Hypnotherapy and the Association for Professional Therapists.

Training: Training in hypnotherapy continues to be very popular in the 1990s. As of 1997, there are some 130 organisations that offer training in hypnotherapy (Berg and O'Sullivan, 1997). However, the quality and length of training vary considerably. Both the NCH and APT have published guidelines of what is considered to be minimum training in clinical hypnosis. However, due to lack of current legislation in this country, there are still those schools training hypnotherapists with suspect means of instruction who also make exaggerated claims as to their recognition (Heap and Dryden, 1991).

Since the early 1990s two UK Universities have sanctioned degree programmes in hypnosis. The Universities are Sheffield and University College London. However, these courses are restricted to those who are either medically, dentally, or psychologically qualified (Centre for Psychotherapeutic Studies, 1997).

Quality Assurance: This issue has become far more important in the 1990s than in the last four decades and has affected the entire profession. No longer is it enough to claim that you are the best; a training provider must prove his or her merits. One of the most effective ways of doing this is to participate in an accreditation programme and these vary widely. City and Guilds have been employed to accredit a training programme in hypnotherapy and counselling (UK Training College, 1997). There is a profession-wide programme of trying to secure National Vocational Qualifications for hypnotherapy (Care Sector Consortium, 1997). Still other providers have entered into negotiations with universities. The most common form of accreditation at this time is through one or more of the independent associations: NCH, APT, ICM, etc. There are still a significant number of training organisations that have no form of external accreditation for their courses (Berg and O'Sullivan, 1997).

With training providers attempting a variety of ways of accreditation, it would appear that the single best way to ensure quality is a properly accountable registration scheme.

"The maintenance of a single register of suitably qualified practitioners, which is accessible to the public, provides the greatest safeguard against possible harm to the individual." (BMA, 1993, pp.130 - 131)

Politics and Culture: In the 1990s there seems to be a greater feeling among practitioners, both medical, psychological and lay, that there is a need for co-operation.

The publication of the 1993 BMA report on complementary medicine seems to have been a watershed in internal and external relationships.

The general public still seek out the services of hypnotherapists, perhaps in greater numbers than at any time before. They will expect the profession of hypnotherapy to put its own house in order, so that people can consult a hypnotherapist in the knowledge that the person they are seeing is competent. It seems likely that any form of restrictive legislation will not find enough interest in parliament (Morgan, 1995).

"...there would be no interest anywhere in introducing such legislation until and unless adequate evidence is offered that abuse does take place and the public does suffer." (Heap & Dryden, 1991, p.193)

In 1997 the Prince of Wales, a long-time advocate of complementary medicine, made a speech for the Kings Fund Trust that has serious implications for the field of hypnotherapy. This speech was directed at having an integrated health service which embraces the best of both orthodox and complementary medicine. It was determined in the report that was conducted by Exeter University (which was also presented at the time of this speech) that hypnotherapy had a long way to go because of the variety of associations and training schools offering hypnotherapy and training (Mills and Peacock, 1997). It would seem that the views of Heap and Dryden regarding the lack of interest in legislative issues regarding hypnotherapy would no longer be the case.

To summarise, there has been a growth in the appearance of professional associations regarding hypnotherapy practitioners. This growth was slow during the 1950s and 60s. In the 1970s there was a gradual increase of bodies, while the 1980s saw a huge surge in new bodies, which still carries on, although not quite to the same extent in the 1990s. Training has developed from solely physicians being trained in hypnosis to the situation today, which is that many mental health professionals and lay people are being trained in the clinical applications of hypnosis. Quality assurance developed rather slowly over the first three decades we looked at. In the 1980s it became more relevant. While in the 1990s there has been a general acceptance from trainers that some form of quality assurance would be desirable, the form this would take is still highly debated.

Finally, hypnotherapy seems to have mirrored the political and social norms of the day. In the 1950s and 60s a more conservative approach was in existence, in that only physicians and dentists practised hypnosis, while things became more flexible in the 1970s. With the desire to let market forces rule the decision-making processes of the 1980s, many associations and training establishments competed for the interested student and practitioner of hypnosis. Now in the 1990s we see a more integrated and more co-operative stance both in society and in hypnosis practice and training.

Chapter 3

Context Of Study

After reviewing the background issues surrounding the study of clinical hypnosis in the UK, the need for this paper is apparent. As a Clinical Hypnotherapist and Trainer with some 8 years of experience in the field, I believe that the present confusion is unhelpful. With so many differing views toward training, it is necessary to look at any similarities in training offered by a number of sources in both the private and public sectors.

In a recent report on an integrated healthcare system in the next five years, it was made clear that the need for research in the field of complementary medicine is essential to develop it into a recognised profession (Mills and Peacock, 1997). I believe that this paper will be useful in the wider debate about professionalism in various healthcare providers that are not currently regulated by statute.

The attitude of the medical profession to lay practitioners of hypnosis has swung from being overtly hostile to that of acceptance, so long as these practitioners can prove competence (BMA, 1993). There is some need for the profession to prove itself, and it is my view that the best way to do this is to look at our training and see if there is some way that we can harmonise courses so that a diploma in hypnotherapy means more or less the same thing at whatever training organisation one attends. History shows that most professions have, in the early stages of their existence, gone through a similar scenario as hypnotherapy is going through now.

It is my hope that, through my research, both training providers and medical and psychological practitioners will see that we have more in common that we share than we have differences.

Many physicians now refer patients to lay hypnosis practitioners, but they do so usually on personal recommendation. It is difficult for medical or psychological practitioners to assess the training a person has received. Though this paper is not about validating any one provider of hypnosis training, it is about the development of training in clinical hypnosis in the UK in the past 26 years and then, taking this information forward into the future and determining where the training of clinical hypnotherapy needs, in my opinion, to go.

Chapter 4

Methodology

Research: The first research that was undertaken was of a scientific base. A point was reached when a researcher could not control all of the variables. This was most especially relevant to the study of human behaviour. This began the evolution of social research which was more qualitative than quantitative.

"Educational research takes a bewildering array of forms, ranging across a variety of topics and employing diverse types of data and techniques of analysis." (Gomm and Woods, 1993, p.ix)

Educational research is primarily interested in challenging the status quo. The difference between the two is that "quantitative" has to do with numbers and figures while "qualitative" has to do with meaning and understandings (Dey, 1993). I chose the latter, as my field of study is less to do with figures and more to do with understanding attitudes and prejudices, although figures will play a part in this paper.

There are several types of research. One is the use of surveys. A researcher can elicit information either from an identified population or from a specified group of people he or she knows. The advantage to this approach is that it gives breadth, generalised ability and descriptive power. However, shallow coverage, bias and insensitivity may be the result. To look into a very limited number of examples of a particular situation, using a variety of ways of collecting data, is known as the case study approach. This is advantageous because the context of the research is familiar. It copes with complexity and can produce intelligible findings.

The case study approach also allows a number of other methods to be adopted and can provide interpretations of similar cases. However, R.K. Yin states that case study is not a method in its own right (Yin, 1989). Also the case history approach can lack scientific vigour, and the participant observer may produce bias.

Approaches to Data Collection

Types of Qualitative Research To Be Used in This Study:
I have opted to combine a documentary search with a questionnaire of convenience as the two main approaches in this study. There were several reasons for this choice.

Documentary Research:
The main one is that, through documentary research, the full responsibility of data collection lies solely with me. I believe that, through analysis of the relevant documents, articles and prospectuses, I can write an informed and well-balanced view of the current situation regarding standards of training in hypnotherapy.

Hypnotherapy Training

I also intend to exploit IT, such as the internet, which will be used in locating more data for my study. I have considered, and will continue to consider, as I see the development of my research, adding other research tools to this method as appropriate.

There are disadvantages to documentary research. The main one is that authenticity context and completeness may be problematic.

Questionnaires/Sampling:
I intend to send questionnaires to leading trainers in the field of clinical hypnotherapy to help evaluate the desirability and necessity of core standards of training.

Because of the present situation with regard to the field of clinical hypnotherapy, it was important to emphasise in my first contact with trainers that this paper was not designed to assess quality (Appendix 2).

My sole purpose was to learn all I could about the training and to discover what aspects there were in common in the training between lay or professional hypnotherapy courses and those being offered to medical and psychological practitioners. I compiled my list of training providers by using the latest publication on hypnotherapy training.

There are currently some one-hundred-and-thirty organisations offering training in clinical hypnotherapy (Berg and O'Sullivan, 1997). I chose my sample by the information provided in this text as well as those who have a reasonably high advertising profile in the complementary medical press.

I chose fifty schools based on length of training provided. This was broken down as follows: ten who provided a one-year part-time training, ten from a two to three-year part-time training, and five from a six-month or less part-time training. I also analysed the years a school has been established. In my original request for prospectuses, ten were to schools established for less than five years and fifteen to schools established for more than five years. Geographical locations played a part in the initial short list. Of the schools contacted, fifteen schools were in or near a major metropolitan area and ten were from other areas. The profile of the organisation with regard to advertising was also considered. Those training bodies that advertise in the main health magazines (e.g. *Here's Health* and *Men's Health*) and professional journals (e.g. *Nursing Times* and *Psychologist*) were contacted. Finally, having graduates still in the field was my final criterion for the initial short list. I reviewed five years of Yellow Pages in Manchester Central and South to see which practitioners were in a minimum of three consecutive editions, and I then enquired where they trained, and approached their schools. In my letter requesting information, I made it clear why I requested their materials, what I intended to do with it, and my own interest as a training provider myself (Appendix 2).

Once I read through the prospectuses I received, I was able to short-list five organisations to write to for further information in the form of a questionnaire. The short-list was derived from the location of the school, the methods of training, and their accreditation system. Additionally, I knew in advance that three of the five organisations would reply to my questionnaire on account of my having personal contacts with them.

Through the information thus received I was able to construct a ten-question questionnaire (Appendix 3). The questions were designed so that I would hopefully get a response from all of the trainers I approached, and also not offend any of them by imposing my particular political views on them.

Analysis (General): There is an obvious need for analysis of data in all research projects. It is this analysis that allows one to draw conclusions from what is found by research. Data for analysis can consist of variables that can be altered and those which are basically unalterable (Gomm and Woods, 1993).

Analysis (Specific to Study): As stated earlier, I have opted to have documentary research form a significant part of this paper. I had to determine which organisations to contact. My method in doing this was to take a broad sample of the varieties of ideology and training methods that are prevalent in hypnosis training today.

By doing this, I found it relatively easy to ascertain the commonalities and differences in the training. For example, there seemed to be a commonality in the number of contact hours used for the main part-time courses. By this I mean that, for all three-year courses, the contact time was similar and, for all the one-year courses, again their hours were similar to each other. It was made clear in the prospectuses.

I received data as to their political views, their entry requirements and, to a lesser degree, what was actually covered in their training courses.

Validity, Reliability and Relevance: The need for validity and reliability in a study is essential when one is researching subjects related to one's own profession. It is important to be aware that a research paper should not be written in order to prove or disprove any particular idea. It is essential to keep an open mind and to accept any finding as being valid, whether or not it is what you thought you would find at the outset of the research. The most reliable way to keep a paper valid is to follow a set of steps and stick to them, even if something goes wrong (Bell, 1997).

Keeping the paper reliable can be done by keeping the sources reliable. I have, wherever possible, checked the information provided in the materials I collected. This was most important with regard to the training prospectuses I received. Prospectuses serve two purposes. One is to inform the potential student of the courses, teaching methods, and qualification the trainer provides. From my analysis and cross-referencing of the prospectuses I received, the claims made were, in the main, accurate.

A prospectus also serves as an advertising vehicle. In other words its function is to get a student to take one particular course of study rather than another. The research I have done has shown this to be less reliable than the course details. In some cases the claims made in the prospectuses were factually incorrect and misleading.

This is not intended as a criticism of any particular course or college, but I had to determine whether, if a prospectus had incorrect or misleading information, it would then be a reliable source of information. I determined that using prospectuses with false or misleading information would undermine the validity and reliability of this paper.

Chapter 5

Findings: Data/Analysis

Documentary Findings:
To the fifty requests for training materials and prospectuses (Appendix 2), I received fourteen replies. Of those, thirteen were either favourable or simply the information I requested. The one that was not favourable I did not follow up, because I felt that there was no point in pursuing my request with someone who seemed, according to his letter, overtly hostile to my request. Based on the above figures, I had a twenty-eight percent response rate. With hindsight, it would appear that my original hopes of seventy-five percent or more response were unrealistic. It is important to note that both the British Society of Medical and Dental Hypnosis and the British Society of Experimental and Clinical Hypnosis had replied. When I originally did my mailing, I did not think that either would reply.

The question that needs to be asked at this stage is, "Why did seventy-two percent of the letters I sent go unanswered?" I am not sure that this should necessarily be interpreted as lack of interest in the research I was conducting.

The results of the documentary survey findings are as follows:

Hours of Training: In the fourteen replies I received, the majority of the training providers gave some indication of the amount of classroom contact that was required for completion of the course. However, the actual hours of live classroom contact, clinical supervision and directed independent study varied widely.

For example, the hours of the two schools that have UKCP validation were laid out completely and unambiguously. There were set criteria for clinical supervision, including qualifications of approved clinical supervisors. These schools had the requirements for directed independent study carefully laid out, so that the prospective student was well informed as to what would be expected in order to attain that school's qualification.

Of the other schools that replied, all but one provided information on course hours. The one that did not was a correspondence college and had no classroom requirement. Only five colleges that responded had published any details relating to supervision. Finally, twelve out of fourteen specified what was required with regard to independent study.

The issue of training hours also came up in the number of years of study that was required to complete the course. Thirteen out of the fourteen responses had set time frames to complete the course. UKCP accredited schools, at the time of publication, had a three-year course duration. This meant that these training providers incorporated extra modules and clinical supervision to make up this time. The remaining schools simply referred to their training as "part-time" and did not make any reference to the number of years a course would take.

Analysis: There is a vast gulf between training providers as to the length of a course and what components make up the total time one studies on a course.

Entry Requirements: Of the schools that replied, thirteen out of fourteen provided some entry criteria that must be met to study at the school. However, the published criteria in many cases appeared woolly. All of the thirteen schools that had published entry requirements named the preferred entry qualification as being a degree in psychology, medicine or social science. If one did not have a degree, a professional qualification in a helping profession such as nursing was also considered to be acceptable. In the published guidelines for entry requirements for the above-mentioned schools, all eleven next went on to say that a certain number of 'A' and 'O' level passes would also be acceptable. Finally, these eleven schools had a provision for potential students who did not fit any of the above-mentioned criteria of allowing mature students to enrol, provided that they either attended a personal interview or proved to the course provider that they would be able to reach the required academic standards for that particular course.

The British Society of Experimental and Clinical Hypnosis and the British Society of Medical and Dental Hypnosis were the only two organisations that accepted only graduates in psychology, medicine and dentistry as potential students.

Analysis: It would appear that, based on the above, the criteria for studying clinical hypnosis are fairly flexible and are down to the individual training provider to determine who is an acceptable student.

Comparability of Curriculum: In all of the fourteen prospectuses I received there was some mention about the course curriculum.

One training provider based a training primarily around one school of hypnotherapeutic thought, that of Milton Erickson. The remaining thirteen schools had a more eclectic format. There were similarities in some of the subjects covered by the training providers who responded to my request. The subjects that were universally covered by the other thirteen trainers were: Basic Ericksonian Techniques, Classical Hypnosis, Neuro-Linguistic Programming, History of Hypnosis, the Elman Technique, Pain Control, Post-Hypnotic Suggestions, Ideo-Motor and Ideo-Dynamic Responses, Practice Management, and the study and understanding of Abreactions.

Most of the courses (i.e. twelve out of fourteen), also incorporate counselling skills and psychotherapeutic theory into their programmes. Two out of the fourteen make specific reference to psycho-dynamic theory. Additionally, twelve out of the fourteen courses make specific reference to the use of hypnosis in the treatment of specific conditions like phobia and stress and behavioural modification therapy (e.g. smoking and weight control). Finally, two out of the fourteen give specific training in anatomy and physiology.

Analysis: Based on the above findings, it would appear that there are commonalities in subjects covered. However, there is no mention as to the amount of time spent on each subject, nor what exactly is covered in each topic. This makes analysis for a common curriculum very difficult.

Accreditation: In all of the fourteen prospectuses I received, some form of accreditation was claimed. Two of the training providers are accredited by the United Kingdom Council for Psychotherapy. This is an independent organisation of associations that promote the ethical use of psychotherapy in some eight main forms of psychotherapy, of which Hypno-Psychotherapy is one. Four are accredited by the National Council for Hypnotherapy. This is one of the oldest-established independent organisations for hypnotherapists in the UK, being the organisation that developed from the Hypnotherapy Register of the National Council of Psychotherapists, founded in 1973. One training provider is accredited by the Association for Professional Therapists. This body was founded in 1986 and accredits hypnotherapists and schools. The remaining seven training providers are accredited by associations that are directly linked to their own school. This form of accreditation can mean that only graduates of the parent school can become members of the association or, as in the case of the British Society of Medical and Dental Hypnosis and the British Society of Experimental and Clinical Hypnosis, only certain professionals can become members of the association.

In addition to association accreditation, two training providers have achieved University Accreditation. Two have published that they are in negotiation trying to secure University Validation. Finally, one provider claims accreditation with both City and Guilds and a UK University. However, this provider's links with these organisations are unclear in its prospectus.

Analysis: There seem to be many forms of accreditation that a training provider can attain, but there is no one organisation that all the providers can agree upon to be the accrediting agency for hypnotherapy.

Codes of Ethics: All fourteen training providers who replied to my request for information claim to follow a code of ethics and practice. Two of the providers subscribe to an external code as well as their own. Five subscribe only to an external code, while the other seven follow only a self-published code of ethics. In respect to hypnotherapy, all codes of ethics have the same weight in law, which is very little. This is due to the present situation in which hypnotherapy is not governed by statute, and as such no code of ethics is enforceable in the courts because it takes an Act of Parliament to determine what is good practice. The fact that a code of ethics is not

Hypnotherapy Training

legally enforceable should not detract from the idea that a hypnotherapist or training provider who subscribes to one is endeavouring to protect the public in some way. Codes of ethics at least give the training provider and general public some form of recourse in dealing with an unscrupulous practitioner or training provider.

Analysis: The idea that all of the training providers subscribe to some form of code of ethics should be seen as an attempt to regulate their own activities in the absence of Parliamentary intervention. The necessity of a code of ethics is an idea shared by trainers no matter what their hypnotic orientation may be. Therefore it is a commonality among trainers.

Questionnaire Findings:
I sent out 5 questionnaires of convenience (Appendix 3) to get other views of the situation regarding the training in Clinical Hypnosis in the UK. I took my sampling from organisations that I either knew would reply – Centre Training School of Hypnotherapy and Psychotherapy (CTS), National School of Hypnosis and Psychotherapy (NSHAP), and Washington School of Clinical and Advanced Hypnosis (WSCAH) – or held views different from my own regarding hypnosis training, i.e. the British Society of Experimental and Clinical Hypnosis (BSECH) and the British Society of Medical and Dental Hypnosis (BSMDH). Of the five questionnaires sent out, I received four back representing an eighty percent return rate. The organisations that responded were CTS, NSHAP, WSCAH, and BSECH. I shall set out each question below with the analysis of the information that I received from these bodies.

i.   The state of hypnosis training in the UK is fine and clear to the public:

On this issue, there was great consensus between the respondents with seventy-five percent strongly disagreeing and the remaining twenty-five percent agreeing. What does this tell us? First, it tells us that these bodies feel that the current situation with regard to training in Clinical Hypnosis in the UK is unsatisfactory, despite these organisations having different political and theoretical perspectives. Second, I think this shows that these organisations may be failing in their ability to educate the public more effectively.

There are numerous programmes on radio and television that pertain to hypnosis or hypnotherapy, but very few of these programmes discuss in any great detail what training is required before someone can actually practise hypnosis. Since 1971, there has been talk about the regulation of hypnosis in relation to other forms of psychotherapy and psychology (Appendix 1). However, even with this view many of the hypnotherapy schools training in 1997 were founded after 1971.

Chapter 5: Findings

Analysis: From the data, I have determined that training providers are not painting an accurate picture of what is involved in becoming a hypnotherapist. If this situation persists, the general public will continue to have a distorted view on the complexities of Clinical Hypnosis training and will, in my view, continue to view hypnotherapy with a great degree of suspicion. This suspicion prevents people who could genuinely benefit from hypnosis seeking treatment because, to the public, the letters after a practitioner's name are in many cases completely meaningless (Brookhouse, 1995).

ii. Hypnotherapy practice should be limited to graduates of medicine, dentistry, and/or psychology:

The results from this question seemed to me to be surprising. Seventy-five percent of responses strongly disagreed, while twenty-five percent agreed.

Analysis: The results from the data seemed surprising to me because the response from the British Society for Experimental and Clinical Hypnosis was in the seventy-five percent which strongly disagreed. Their stated policy is to exclude anyone outside this group from learning about hypnosis in order to apply it in a therapeutic setting (Centre for Psychotherapeutic Study, 1997).

iii. Legislation is required to regulate the training and practice of hypnosis:

To this question there was a greater difference of opinion. Fifty percent disagreed with this position, twenty-five percent had no opinion and twenty-five percent agreed with this position. The organisation that has been closely linked with the view of statutory legislation seems to have shifted its view. Both the British Society of Experimental and Clinical Hypnosis and Washington School of Clinical and Advanced Hypnosis disagreed that legislation was required. National School of Hypnosis and Psychotherapy had no opinion. Centre Training School agreed with this position. The views of the Principal of that school regarding legislation are well known in the field.

Analysis: These results may have significant repercussions for the hypnosis training field in general. It would appear that there is no common policy regarding legislation of some kind with regard to Clinical Hypnosis training. This, like the lack of a common core Course of Study with regard to the training of hypnotherapists, leads to the fractionalisation that is now occurring in this profession. It would appear from the replies that the view that legislation would be a quality control measure in its own right has been challenged, and the profession may need to re-evaluate its views on governmental control both here and in the European Union.

iv. Legislation is likely in the next ten years to cover question 3:

The replies for this question were not surprisingly similar to those for the above question.

v. Your training would recognise graduates from other colleges:

The results from this question were split as follows: sixty percent would accept graduates from other colleges, twenty percent would not, and for twenty percent the question was not applicable. This answer really had five replies. Dr Heap, who replied on behalf of the British Society of Experimental and Clinical Hypnosis, answered this question in both his role as President and as Administrator of Clinical Hypnosis Courses at the University of Sheffield.

Analysis: The majority of respondents would accept graduates from other colleges even though no core Course of Study exists. However, the reply from the British Society of Experimental and Clinical Hypnosis specified only three organisations whom they would recognise. These are: The British Society of Medical and Dental Hypnosis, Sheffield University, and University College London.

The other organisations that replied in the affirmative, Centre Training School and Washington School of Clinical and Advanced Hypnosis, did not specify whose graduates they would accept. The only responding organisation that replied that they would not recognise qualifications from other colleges was National School of Hypnosis and Psychotherapy.

Though reasons were not asked for or given for this view, National School of Hypnosis and Psychotherapy is in a unique position, as their training is based almost solely on the teaching of one particular school of hypnotherapeutic thought, the work of Milton H. Erickson, MD (NSHAP, 1997). The other organisations have a more eclectic training and cover a variety of schools of therapeutic thought.

vi. Your training has some form of University validation:

Again, as in the above question, there were in effect five replies. This is for the reason outlined above. The breakdown of the figures were that eighty percent did not have some form of University validation and twenty percent did. Centre Training School, which at present does not have University validation, replied no, but it is currently being negotiated. The twenty percent that does have University validation is actually a course that is run by the University of Sheffield.

Analysis: The information shows that the lack of a core Course of Study has caused difficulty for Universities who may be inclined to validate hypnotherapy courses. With hindsight, I probably would have asked this question differently. I think a more appropriate question would have been "Do you desire University validation?" or "Have you attempted to secure some form of University Validation?" Certainly one respondent, CTS, seems to think that having University validation is desirable. With regard to recognition in the profession of hypnotherapy by the public and the National Health Service, it would appear that University validation will become more and more necessary as a form of quality control.

vii. Voluntary self-regulation would be a useful system of quality control for hypnosis training:

The replies to this question were consistent with the replies given by these organisations to question 4 of this questionnaire. Fifty percent of those who responded said they strongly agreed, twenty-five percent agreed and twenty-five percent disagreed.

Analysis: The one thing that could be said for all of these replies was that some form of quality control was desirable. I draw this conclusion from the respondent that disagreed with self-regulation as a form of quality control, but agreed with the statement about using legislation as a means of quality control. Therefore, the bodies that replied to my questionnaire all felt that some form of quality control would be desirable for Clinical Hypnosis training.

The majority of those who replied also felt that self-regulation would be the most appropriate way to achieve this. The difficulty is that there is no agreement between those who represent mental health and medical professionals and those who represent the lay community as to the best way to implement self-regulation.

viii. NVQs are a positive development for hypnosis training:

The replies to this question were heavily weighted against NVQs. Seventy-five percent disagreed while twenty-five percent had no opinion on the idea that NVQs were a positive step for hypnosis training.

Analysis: The reason for this might be the inference that NVQs are a tradesman's qualification rather than a qualification for professionals.

ix. Correspondence courses provide ample training for hypnotherapists:

This was the only question that received one-hundred percent agreement. All disagreed strongly. This information could be used to determine at least an agreement as to how hypnosis training should be conducted. All of the course providers who replied have a component of their teaching based around some form of home study and they also all have several hundred hours of live contact time between trainer and student.

Analysis: Though these hours vary, it seems clear that the contact time between trainer and student is essential in the effective training of a hypnotherapist. As one of my research objectives was to find commonality between organisations that train lay people as well as mental health professionals in hypnosis, this is one area of the debate that my sampled trainers can agree on. Where there is agreement on one issue, there is the hope of agreement on other issues.

x. Hypnotherapy is a profession in its own right:

The replies to this statement are as follows: fifty percent strongly agree, twenty-five percent agree, and twenty-five percent strongly disagree.

Analysis: These results are consistent with what I expected at the outset of my research. The schools that train both lay and mental health professionals were in agreement with the idea of hypnosis being a profession in its own right, while the organisation that represents mental health professionals solely disagreed. This is entirely consistent with their stated ideas in their prospectuses.

Chapter 6

Discussion

This paper has given me the opportunity to look historically at the situation with regard to the training in Clinical Hypnosis in this country. It is my belief that this research will serve as a jumping-off point for other researchers in this field to ascertain the status of this profession from public and professional perspectives. Most important of all, I believe that this work could become Chapter One in a history of the evolution of the hypnotherapy profession from the early 1950s to today. Of course, this would probably be in the distant future, and my hope would be that future researchers would see this as an important period of development, but by no means an ending in itself.

Sadly, there is a mistrust in this field with regard to sharing information between training bodies (Berg and O'Sullivan, 1997). In fact, there is practically no sharing of information or research between training colleges (BMA, 1993). I can only hope that, after this publication, others involved in this field will take a greater interest in researching the training provided to hypnotherapists in the UK. In some ways, this point is already being looked at through an organisation founded just after the start of this research. The Hypnotherapy Research Society (UK) has set out to:

"...encourage practising hypnotherapists and researchers of clinical hypnosis to publish or write papers of their theories, experiences, and any practical research they have undertaken in the field of clinical hypnosis or hypnotherapy." (HRS, 1997, p.2)

The coming into existence of a research society for hypnotherapy seems like a huge step forward when one compares the level of hypnotic research now with what was done in the early fifties, with this paper beginning to look at those four key points: Growth of Associations, Training, Quality Assurance, and Politics and Culture.

One question that remains unanswered after my research is the issue of National Vocational Qualifications (NVQs) in hypnotherapy. Whilst in many of the prospectuses I received these are mentioned, only one openly embraces them as the way forward (UK College, 1997). It would appear that many trainers are treating NVQs as an inevitability rather than as a step forward (Care Sector Consortium, 1997).

Another issue that I feel is unanswered by my research is why there still remains a hostility between physicians and psychologists with regard to the professional lay hypnotherapist.

Based on the training hours of the main courses approved by the British Society of Experimental and Clinical Hypnosis and the British Society of Medical and Dental Hypnosis and those courses provided for professional lay hypnotherapists, the hours are comparable.

Hypnotherapy Training

Also, the teaching staff in many of the professional lay schools are similar, and in some cases arguably superior, to those on the main medical and psychological courses. Over fifty percent of the professional lay schools that responded have at least one psychologist or physician on their training faculty.

Though one can see that this is not the same as a Medical School education, one must ask, "Would a medical or psychological practitioner endanger his or her reputation by being involved with an unethical or unprofessional training provider?"

I have identified six main barriers to the acceptance of lay hypnotherapy training by the medical and psychological communities. The first is that there is no common curriculum taught by all schools. The diversity of training courses and their content leads to confusion when trying to assess the differences between the skills possessed by the hypnotherapists and other mental health professionals. Many fields that have achieved "recognised status" with the organisations like the BMA or the government went through a period of harmonisation before this recognition was achieved (BMA, 1993). However, this process of harmonisation within hypnotherapy does not seem to have yet begun, based on the prospectuses sent to me.

The second is that entry requirements vary greatly, with some courses requiring a degree for entry, while others have no academic requirement. Many professions accept the idea of a graduate level of entry. This requirement is based on the belief that a university education is a good grounding for a professional to have, even if the undergraduate study bears no relation to the field a graduate chooses to enter (UKCP, 1997). The lack of university interest in the field of hypnotherapy leads me to deduce that it will be some time before undergraduate courses in hypnotherapy will be made available.

The next is that length of course study ranges from one or two weekends of training, to home study courses, to three-year courses with clinical supervision and personal therapy requirements. For many professions outside hypnotherapy, this seems anathema. How can a person train to be a hypnotherapist, on the one hand, in two weekends, while others take three years of part-time study? Ironically, the organisation with the shortest requirement as far as training is concerned is the British Society of Medical and Dental Hypnosis. Their certifying course is two weekends (BSMDH, 1997). However, they explain that the reason for this is that their graduates have all undergone a rigorous medical or dental training. Leaving the BSMDH out of this, most of the other training requirements advertised in the prospectuses I received had a minimum of 1 year part-time training requirement, with additional requirements for Clinical Supervision and In-Service Training, for full qualification (CTS, 1997).

Then there is a range of qualifications offered (e.g. DHP, CHP, Dip CAH, Dip THP) with no way of assessing their relationship with each other or their relationship with higher awards (e.g. University Degrees, HNDs, etc.). For a student this is problematic. If one wants to study to be a osteopath, for example, there is one main qualification so there is less room for being confused (BMA, 1993). In many cases, hypnotherapy qualifications can sound more impressive then perhaps they really are. For example, a correspondence course prospectus I received stated that, upon successful completion of their diploma, one would be eligible to join an "international association" whose membership is the largest in Europe (IAH, 1997).

When one actually looks into this society, it is shown that it is not the largest association in the UK let alone Europe (Mills and Peacock, 1997). Also, the only persons who are eligible for membership of this association are graduates of the school in question. It would be like Manchester University having its own professional society for psychologists open only to graduates of their psychology programme (Brookhouse, 1995). From what I have seen in the prospectuses sent to me, most training organisations are giving a fair definition of their qualifications. However, it is still difficult to determine how they equate with each other and whether one qualification based on the designating letters requires any more or less time of study than another.

This leads to the fifth barrier. Since there is no standard of education for "lay" hypnotherapists, the medical and psychological communities have sought to ban their practice (Heap and Dryden, 1991). There is a case for this, as physicians and psychologists go through a rigorous course of undergraduate and graduate study, but the same cannot be said of the "lay" hypnotherapist. In the literature of the British Society of Experimental and Clinical Hypnosis, it even goes so far as to ban any member from teaching a non-qualified person hypnosis:

"A student shall not give courses involving the teaching of hypnotic techniques to lay individuals. A lay person here is defined as one who is not a member in good standing of a therapeutic or scientific profession. Lectures informing lay individuals about hypnosis are of course admissible providing they do not include demonstrations or didactic material involving the induction of hypnosis." (Centre for Psychotherapeutic Studies, 1997, p.12)

Though many "lay hypnotherapists" are university educated, there are a number of practitioners who have completed only a two-month correspondence course to allow them to practise. Because of the UK's legal tradition of common law, anyone can set themselves up as a hypnotherapist without any training at all. Training colleges that responded to my letter who provide a live part-time training seem to be in agreement on the undesirability of this situation (NSHAP, 1997).

The sixth, and final, barrier has been getting Universities to validate training for "lay" hypnotherapists. Since the mid 1950s the British Medical Association has endorsed the training of hypnosis to physicians. There are currently two degree courses at graduate level for "applied hypnosis" in the UK. These courses are run at University College London and Sheffield University. These courses are restricted to psychologists, physicians and dentists. The "lay" hypnotherapist is left out in the cold. Some practitioners have sought university qualifications from the United States. However, not all of these courses are properly regulated and there has recently been a backlash against US degrees by many "lay" hypnotherapy organisations. This is due to the fact that in the late 1970s and early 1980s "degree mills" advertised and sold many "degrees" to therapists who felt that they needed the prestige of an academic qualification. Some organisations felt that the problem of bogus doctorates in the field of hypnotherapy was so serious that they made provisions and rules relating to the use of the title "Doctor" in professional practice (Appendix 5).

Bearing this in mind, there are a few licensed degree-granting institutions who are authorised by their particular state to award professional degrees rather than academic degrees in hypnosis and hypnotherapy (AIH, 1997). However, in the 1970s and 1980s a UK organisation was awarding degrees so that their graduates could become "Doctors of Hypnosis". This practice ceased with the implementation of the 1988 Education Act which forbade organisations from awarding degrees unless they were chartered to do so by the Department of Education (Morgan, 1994). Though this practice ceased ten years ago, some of these practitioners are still in business and very often continue to use the title "Doctor of Hypnosis", which can cause confusion to the general public and condemnation from the profession of hypnotherapy as well as those outside, such as physicians, dentists and psychologists.

In addition to these six barriers to the acceptance of a unified hypnotherapy training by the medical and psychological communities, I have also identified some bridges to this acceptance. The first is that all of the training colleges subscribe to a code of ethics and practice. Some subscribe solely to an internal code, while others subscribe to both an internal and an external code. There was only one that subscribed solely to an external (WSCAH, 1997). Though these codes are not enforceable by statute at this time, I believe they show that trainers are willing to subscribe to rules and regulations regarding not only their students' behaviour, but also their own behaviour as trainers. This notion that trainers and graduates subscribe to a code of ethical behaviour seems to directly contradict views of some professionals outside the lay field of hypnosis.

"Hypnotic techniques should not be practised by anyone other than licensed medical or mental health professionals, as lay persons who practise hypnosis do not subscribe to an enforceable code of ethics and practice." (Ewing, 1996, Seminar Bury Postgraduate Medical Centre)

Also, lay training colleges consult with medical and psychological practitioners and, in some cases, have a physician, psychiatrist, and/or psychologist on their teaching staff. One of the training providers who sent in their prospectus has a Professor of Primary Health Care on its staff (CTS, 1997).

The idea of having a lecturer who is a licensed health professional is not, nor is it implied, the same as attending a training in psychology or medicine, although it does, in my view, show a desire to understand more about these professions in relation to professional practice. It also can be implied that the schools who do have such professionals on their staff are prepared to alter any controversial parts of their training in line with what is expected by the medical and psychological communities.

Disclosure also seems to be a possible bridge for the lay hypnotherapy community and the medical and psychological communities. Only one of the twelve prospectuses I received did not list their teaching and administrative staff. Some of these prospectuses went into great detail with regard to their staff's experience and qualifications.

Many of the key members of staff at the schools who provided prospectuses are graduates in psychology or social science, and there are even those who hold master's degrees and doctorates in these subjects, which I believe shows, based on the arguments of the medical and psychological communities, a high level of quality with regard to the dissemination of the theories around the therapeutic applications of hypnosis.

Regarding the issue of National Occupational Standards, five out of the twelve training colleges who provided prospectuses are ready for any future implementation of National Vocational Qualifications (NVQs) for hypnotherapy. These five schools have key members of their training and assessment staff who are qualified to D32 and D33 levels of NVQ assessors.

Based on this it seems that, like the medical and psychological professions, the field of hypnotherapy is looking for ways to continue to improve on the qualifications offered and their validity both in the UK and the world outside (CSC, 1997).

The area of contra-indications can also be seen as a possible bridge between psychological and medical practitioners and lay hypnotherapists. In all, twelve of the prospectuses received made some reference to the study of contra-indications in hypnotherapy. It is the study of what a particular discipline is not useful for, which is a necessity for professional competence. The need to practise within one's scope of competence is made clear in all of the training materials received.

I believe that this is where the medical and psychological practitioners on a school's staff can really be useful. Having a physician or psychologist tell a student, or even a graduate, that a certain case is either inside or outside their scope of competence to treat gives that student peace of mind that the clients that they are seeing are those they are qualified to treat. I think the research shows that the argument that only physicians are qualified to make a determination as to what constitutes a contra indication does not hold (BSMDH, 1997).

The final bridge I have found through my documentary research is the surprising amount of physicians and psychologists who have gone through lay hypnosis colleges' training courses. I determined this by several means of checking. One is the number of medical and psychological staff who trained at the college in question. Another method of checking this is through testimonials that some providers include with their prospectus. Some of these professionals rate the courses that they attended as high as the courses they attended at university.

"I am writing to you on my successful completion of the course and to thank you for the professional knowledge, experience and advice which the course has given me. As a holder of a doctorate in the area of Jungian psychology from Corpus Christi College, Oxford, and as a former university lecturer with experience of accredited Honours degree course design, I found the course to be enjoyable, well designed, and presented, and to equip me with all the skills necessary to set up a practice." (Woodbury Counselling, 1996, Dr Chris Harvey Testimonial, p.10)

As I stated earlier in this paper, the difficulty in using prospectuses for research purposes is that they not only convey what a course provider is offering, but are also an advertising tool. That being the case, it would seem that a professional psychologist is not likely to endanger his reputation by making false or misleading statements regarding a training course provider. So, I believe, using testimonials as evidence in this context is safe and valid.

With all of these advancements in the training of clinical hypnotherapists I think that the quote from David Waxman in 1989 is no longer valid.

"First of all it must be understood that there is no authentic degree or diploma in hypnosis issued either in Great Britain or any other country in the world. There is no such person as a 'qualified' hypnotherapist..." (Waxman, 1989, p.480)

Another issue that was raised by my research is the importance of Europe. It is ironic to me that this very important issue did not even exist back in the 1950s, the first period I researched in order to compile this paper.

The importance of this issue relates to the implementation of legislation as a form of quality control. My research shows that the majority of responses to my questionnaire not only do not see legislation as effective quality control, but also that these bodies do not believe any statutory legislation will be imposed on the profession from the governments of either the United Kingdom or the European Community.

This is a major shift in attitudes from even six years ago. In a letter sent to many hypnotherapists from the Institute for Complementary Medicine, it was implied that failure to join its British Register of Complementary Practitioners would result in a practitioner's loss of the right to practise either here in the UK or in the EC.

The Institute for Complementary Medicine was not alone in its concerns that legislation from the European Union was inevitable and likely to be introduced soon after the Common Market converged in 1992. However, it is now believed by many who have dealings with the European Parliament lobbying for psychotherapy as a profession that this legislation will probably never come (Van Durzen, 1998). I believe that with the spectre of legislation now exorcised, the field of hypnotherapy can now begin to implement voluntary controls which will protect the public with regard to the training in, and practice of, clinical hypnosis.

Chapter 6: Discussion

Another important aspect of this research was to look at the rise of a few independent registers of practitioners of hypnosis who now serve a course accreditation role. The field of hypnotherapy has a real chance for recognition by the medical and psychological communities and, after that, the Government, through the Department of Health and possibly the Department of Education and Employment, through the advent of National Occupational Standards.

And finally, there are a few schools who are seeking University validation. Some training providers are resistant to the idea of National Occupational Standards and would prefer the idea of University validation. One of the training providers who submitted their prospectus for my research has gone into partnership with De Montfort University.

By that, the school offers a qualification in conjunction with the university (UK College, 1997). Another has arranged that its course is worth a one credit toward a degree at the Open University (Woodbury Counselling, 1997). There is also a training course provider that is attempting to get its entire diploma course validated by a UK university at master's degree level (CTS, 1997). I observed in the prospectuses I received that many training providers made more of an issue about who was presenting the training rather than the information and skills that students will be expected to learn on a particular course of study. Once this difficulty can be overcome, I believe that training in clinical hypnosis will be more appealing for Universities to consider.

Looking back on the research I have conducted, it is my hope that this paper will begin to build bridges with regard to the issues around training. I suppose the question that I am left with is, "Why has it taken so long for the profession to start really looking at the issues around training?" The fact that so few of the trainers I approached actually felt that it was in their interest to help me by simply sending a prospectus indicates to me a lack of trust or interest which leads to misunderstanding and friction between factions within the hypnosis community.

It is my sincere hope that this paper will generate the interest of my colleagues, who will pick up my work and expand it further. It is a requirement, in my view, that for hypnotherapy to be taken seriously as a profession we need not only our techniques critiqued, which has been done in countless scientific and professional journals, but also to have the way we practise and train to undergo the same scrutiny. Through this scrutiny, I believe that the profession can grow and become recognised by both other professionals and the general public alike. It is my view that training is the most important issue with regard to any profession.

If the profession has a properly regulated and scrutinised course of training, the public can feel secure in seeking a hypnotherapist to help them to alleviate the conditions that hypnotherapy has been scientifically proven to treat. So the issue with regard to training has not been resolved by this research, but this paper has, I believe, given other professionals an opportunity to look closer at the way hypnotherapists are trained, and hopefully to find new and better ways of doing this. If this should happen, then I will have truly achieved something special.

Since my research began, one state in the US has passed a registration act for hypnotherapists while another state is in the process of passing a licensure law relating to hypnotherapists (HB 1508 FN, 1997). These events are important to this study because these laws are making hypnotherapy a profession in the US just like any other mental health professional. These laws set minimum training standards as well as the scope of practice for hypnotherapists.

These events are viewed in the hypnotherapy community in the US as being historic (Damon, 1997). Though we are not in a similar legislative position as American hypnotherapists, it does show that hypnotherapy is now being viewed as a discrete profession in its own right. The repercussions of this cannot help but influence views in the UK.

Chapter 7

The American Experience

This is a newly created chapter for this latest edition. I believe that it will prove useful as a case history of the legislative and training issues from what is arguably the nearest society to the United Kingdom. Having links in the United States has undoubtedly been of assistance in the compliation of this chapter, and my heartfelt thanks go to all those American Hypnotherapists and Organisations who helped me.

Though the situation in legal terms differs widely between the United States and the United Kingdom, the aspects of training are surprisingly similar. In the US, there are three main professional societies for clinical hypnosis for medical and psychological practitioners: these are the American Society of Clinical Hypnosis, the International Society of Experimental and Clinical Hypnosis and the Milton H Erickson Foundation. In the "professional" hypnotherapy community there are three main bodies: The American Council for Hypnotist Examiners, the American Board of Hypnotherapy, and the National Guild of Hypnotists.

The American Council of Hypnotist Examiners is supported by 40+ affiliated training institutes which offer training at Master Hypnotist, Hypnotherapist, and Clinical Hypnotherapist level. Hypnotherapists who have not trained at an ACHE approved institute can still achieve certification by passing the ACHE's examination. The ACHE was founded in 1980 by Gil Boyne of the Hypnotism Training Institute of Los Angeles. The Hypnotism Training Institute was founded in 1956. In 1976 it was the first state licensed hypnotherapy institute in the United States. (ACHE 1998). The American Council of Hypnotist Examiners was instrumental in achieving the first hypnotherapy certification law in the United States. The Indiana Certification Act requires that hypnotherapists in that state have completed 200 hours of instruction and training at a state approved hypnotherapy school. There is also a 150-hour supervision requirement. The Certification as a hypnotherapist is granted by the Indiana Hypnotist Committee which is a committee of the Indiana Medical Board. At the time of publication, the Indiana Hypnotist Committee is in the process of instituting a grandparenting period for those hypnotherapists who have completed 300 hours of training at a state approved school of hypnotherapy. Thus experienced hypnotherapists will not have to re-train provided they have achieved the requisite number of hours. The ACHE's activities are not just limited to the United States. Mr. Boyne, the ACHE Executive Director has co-founded the British Council of Hypnotist Examiners as well as helping to secure practitioner rights in Australia, for non-psychologists.

The American Board of Hypnotherapy is supported by the State Approved Degree Granting Institution, the American Institute of Hypnotherapy, American Pacific University, a University chartered in Hawaii, and 300+ affiliated institutes. The ABH also has certified instructors who operate internationally, who provide the requisite tuition for admission as a Certified and Registered Hypnotherapist. Currently the requirement for Certification is 100+ hours. This however, is likely to change as a result of possible state regulations and other professional considerations. The American Board of Hypnotherapy was founded by Dr. A.M Krasner and Dr. Bruce Koloski in 1982. Its original name was the California Council of Hypnotherapists. In 1987, the name was changed to the American Board of Hypnotherapy to better represent the growth of the Membership to a national level (ABH 1998). However, in 1998 Dr. Krasner retired, with Dr. Everett W. (Tad) James taking over as both President of the American Institute and as Executive Director of the American Board of Hypnotherapy. The American Institute of Hypnotherapy is a California State approved degree granting institution founded in 1982.

It offers Bachelor and Doctor of Clinical Hypnotherapy degree courses, primarily through directed independent study, but there is also a practicum requirement (AIH 1997). The American Pacific University also awards Bachelor and Doctor of Clinical Hypnotherapy degrees, as well as a PhD in Hypnotherapy, the PhD and PsyD in Clinical Psychology and a Doctor of Esoteric Studies Degree and a PhD in Esoteric Studies (APU 1999). According to my research, of the 109 presentations at the National Guild of Hypnotists' Convention in 1995, 17% of the presenters had already obtained or were working towards an AIH Degree (NGH 1995). Though the AIH degrees are state approved and the APU degrees are internationally accredited, there are some in the hypnotherapeutic community who are opposed to them. As the Author, I must declare an interest as I am a graduate of the AIH programme and found it to be an excellent supplement to my already considerable training in clinical hypnosis. I am also a faculty member of the American Pacific University and believe degrees such as those offered at AIH and APU are the future of continuing professional development for hypnotherapists worldwide.

The National Guild of Hypnotists is supported by a network of international Certified Instructors of Hypnotherapy. The National Guild was founded in 1951 as a not-for-profit educational organisation. This makes it the oldest society for hypnotherapy in the US. The National Guild has formed a Trade Union, The National Federation of Hypnotists 104, which is its legislative arm. In the past it would appear that several smaller hypnosis organisations merged with the National Guild, organisations such as the International Hypnological Association, the Hypnosis Educational Council International, the National Board of Hypnosis Education and Certification, and the National Association of Clergy Hypnotherapists. The National Guild of Hypnotists produces both the Journal of Hypnotism and Hypno-Gram. These publications cover the whole field of hypnosis and hypnotherapeutic techniques. The National Guild's certification standards

Chapter 7: The American Experience

require 100 hours of training for basic certification as a hypnotherapist with an additional 50 hours for advanced certification. In 1999, the NGH is planning to offer board certification for therapists who desire more training than either of these two levels of certification (NGH 1998).

Though the above are the main players in US professional hypnotherapy (by "professional" I mean practitioners who solely practice hypnotherapy as a career), there are also a number of smaller bodies. Many of these are "in house" certifying bodies. The term "in house" means that primarily the members of these bodies have graduated from a school of hypnosis that is owned by or is in some other way linked to the professional society. This is not to say that in house bodies are good, bad or indifferent. As with so many aspects of hypnotherapy, the only way to make an evaluation is through experience. Many of these bodies are represented in an organisation called the Council of Professional Hypnosis Organisations or (COPHO). Many people feel that COPHO started with the right ideals, but in reality its potential has never been realised. Part of the difficulty with the organisation, as I see it, is that 2 of the 3 main bodies previously discussed are members. These two organisations have, by far, the majority of practitioners represented. But neither of these two organisations can "veto" the will of the smaller bodies. In many respects this causes a problem, as larger organisations with representatives in more than a couple of States may have a better view of the big picture than those smaller bodies who may have only a special interest in their own survival or well-being.

Hypnotherapy continues to grow and develop in the United States, and it seems from my research that the overt hostility between the psychological/medical/dental communities and the professional hypnotherapist has subsided. As in the UK, American organisations and practitioners simply want to be able to practise hypnotherapy in a free, open and honest way. Although there are those who would undermine the credibility of the profession, I believe that the systems are well in place to ensure that the profession will continue to evolve and become more recognised in America as well as here in the UK.

Chapter 8

The UK Confederation of Hypnotherapy Organisations

On the 27th of June 1998, an historic meeting took place that wa attended by the majority of hypnotherapy organisations in the UK. The meeting was hosted by the National Council for Hypnotherapy, and chaired by the author. The purpose of the meeting was to ascertain whether it would be desirable for the main organisations in UK Hypnotherapy to co-operate to form a National Umbrella Body. A unanimous decision was made to form the UK Confederation of Hypnotherapy Organisations. Its prime purpose is to enable hypnotherapy organisations to begin the arduous process of recognised self regulation. The UKCHO has already had a favourable meeting with the newly established Health Care National Training Organisation for Health Care. Its remit is to ensure that occupational standards are implemented throughout the health care community.

The UKCHO was formed with the idea of equality in mind and that organisations should not be hostile to other member organisations (UKCHO 1998). It has been this atmosphere of distrust and hostility that has tended to keep professional hypnosis in a rut over the past 10 or more years (Brookhouse 1998).

I felt that it would be a useful resource in this updated volume to add the names and status of the various members of the UKCHO. They are alphabetised into the following sections: Training Members, Registering/Accrediting, Registering, and Friend. I hope that readers will find this helpful when consulting a hypnotherapist or contemplating becoming a trained hypnotherapist.

The UK Confederation of Hypnotherapy Organisations
Chairman: Dr. Shaun Brookhouse
Vice-Chair: Anne Billings
Secretary: Martin Armstrong Prior
Suite 401, 302 Regent Street
London, W1R 6HH
Tel: 0870 6070422 (National Rate Call)
Fax: On Request
E:Mail: UKCHO@hypnotherapy.demon.co.uk

UKCHO Membership List
As of 9th November 1998

Training:

Academy of Curative Hypnotherapists Ltd.
Contact Person: Simon Kilner
16 Station Rd
Cheadle Hulme
Stockport
SK8 5AE
Tel: 0161 485 4009
Fax: 0161 485 4009
E:Mail: ach@zen.co.uk

Atkinson Ball College of Hypnotherapy and Hypnohealing
Contact Person: Cherith Powell
PO Box 70
Southport
PR8 3JB
Tel: 01704 576285
Fax: 01704 576285

Avalon Hypnotherapy
Contact Person: Dr. Christopher Forester
3 Benedict Street
Glastonbury
Somerset
BA6 9NE

British Association of Therapeutical Hypnotists
Contact Person: Jean Murton
The Beamont Centre
46 Belmont Rd
Ramsgate
Kent CT11 7QG
Tel: 01843 587929
Fax: 01843 587830

Centre Training International School of Hypnotherapy & Psychotherapy
Contact Person: Sue Washington
145 Chapel Lane
Longton
Preston
Lancs PR4 5NA
Tel: 01772 617663
Fax: 01772 614211

The Dominic Beirne School of Hypnosis and Psychotherapy
Contact Person: Dominic Beirne
The Stables
Welcome Rd
Stratford upon Avon
CV37 6UJ
Tel: 01787 261620

East Midlands Psychotherapy Training
189 Uppingham Rd
Contact Person: Martin Armstrong-Prior
Leicester
LE5 4BQ
Tel: 0116 2764911
Fax: 0116 2764911

Innervisions Complementary Health Centre
Contact Person: Brian Glenn
421 Hessle Rd
Hull
HU3 4EH
Tel: 01482 322203

Institute of Clinical Hypnosis
Contact Person: Robert Dalrymple
28 Tantallon Rd
London
SW12 8DG
Tel: 0181 675 1598
Fax: 0181 673 2230
E:Mail: ICHypnosis@aol.com

Irish School of Ethical and Analytical Hypnotherapy
Contact Person: Dr. Joseph Keeney
Therapy House
6 Tuckey St
Cork City
Ireland
Tel: 00353 (0) 21 273575

Joe Leo Associates Ltd. t/a Newcastle Centre for Hypnotherapy & Psychotherapy
Contact Person: J.L. McAnelly
9 Bedeburn Rd
Wmorlton Grange
Westermore
Newcastle upon Tyne NE5 4JL
Tel: 0191 286 1161
Fax: 0191 286 4090
E:Mail: JoeLeo@BTInternet.com

London School of Eclectic Hypnotherapy & Psychotherapy
Contact Person Lyn Buncher
808a High Rd
Finchley
London
N12 9QU
Tel: 0181 446 2210
Fax: 0181 446 2210
E:Mail: eclectic.therapy@btinternet.com

Mind Train
Contact Person: Robert Chambers
The Island Health Centre
145 East Ferry Rd
London E14 3BQ
Tel: 0181 505 5078

Mindworks Therapy Training
Contact Person: Pat Doohan
46 Highbury Avenue
Bulwell
Nottingham NG6 9DB
Tel: 0115 9278791
Fax: 0115 9136697

National Centre for Therapeutic Studies
Contact Person: Philip Clark
24 Byron Mews, off Fleet Rd
Hampstead
London
NW3 2NQ
Tel: 0171 428 0568
Fax: 0171 284 2199

National School of Hypnosis and Psychotherapy
Contact Person: Polly Roche Pengelly
28 Finsbury Park Rd
London
N4 2JX
Tel: 0171 226 6963
Fax: 0171 226 6963

The College of Transformational Therapy
Contact Person: R.S. Piggott
212 Porchester Rd
Nottingham
NG3 6LH
Tel: 0115 9504511
E:Mail: subcon@which.net

Twa Acres Natural Therapy Centre
Contact Person: Vicki Watson
Westwood Londge
Carsie Blairgowrie
Perthshire PH10 6QW
Tel: 01250 874384
Fax: 01250 873598
E:Mail: twaacresntc@compuserve.com

Washington School of Clinical and Advanced Hypnosis
Contact Person: Dr. Shaun Brookhouse
Richmael House 25 Edge Lane
Chorlton
Manchester M21 9JH
Tel: 0161 882 0400
Fax: 0161 882 0376
E:Mail: WSCAH@hypnotherapy.demon.co.uk
Internet: http://www.hypnotherapy.demon.co.uk

Woodbury Counselling Ltd
Contact Person: Deborah Greaves
Woodbury House
Woodchurch Rd
Tenterden
Kent TN30 7AE
Tel: 01580 763286
Fax: 01580 766648

Registering & Accrediting:

Hypnotherapy Research Society (UK)
Contact Person: Dr. Christopher Forester
Milfield Business Centre
Stone in Oxney
Kent TN30 7JL
Tel: 0181 428 5393

National Association of Counsellors, Hypnotherapists & Psychotherapists
Contact Person: Anne Billings
Aberystwyth
Dyfed
SY23 4EY
Tel: 01974 241376
Fax: 01974 241795

National Council for Hypnotherapy
Contact Person: William Broom
Hazelwood
Broadmead
Sway, Hants
SO41 6DH
Tel: 01590 683770
Fax: 01590 683770

Registering:

Association of Professional Therapists
Contact Person: Douglas Simmons
Katepwa Hse
Ashfield Park Business Centre
Ross on Wye
Herefordshire HR9 5AX
Tel: 01989 764905
Fax: 01989 567676

Chapter 8: The UK Confederation of Hypnotherapy Organisations

Association of Qualified Curative Hypnotherapists
Contact Person: Judy Hopkins
10 Balaclave Rd
Kings Heath
Birmingham
B14 7SG
Tel: 0121 441 1775

British Institute of Hypnotherapy
Contact Person: P.C. Lawrence
12 Heycroft Rd
Eastwood
Essex
SS9 5SW
Tel/Fax: 01702 524484
E:Mail: bih@globalnet.co.uk

Lambourn Court Ltd
Contact Person M. Lloyd-Owen
PO Box 328
Huddersfield
W. Yorks
HD4 3YP
Tel: 01484 461462

Friend of UKCHO

James Braid Society
Contact Person: Leila Hart
7 Radnor Mews
London
W2 2SA
Tel: 0171 402 4311
Fax: 0171 402 5679

Epilogue

When this research was originally undertaken, I realised that the work would probably leave more questions than answers. Due to the limitations of its scope, it was impossible to cover all of the items that had to do with training in clinical hypnosis. It is my hope that this book will serve as a jumping-off point for other researchers interested in the development of this burgeoning profession. I hope that you, the reader, gained an insight into the training position within clinical hypnosis, and that it served to help you in your search for information. If you are looking to train in hypnotherapy, I wish you well, for despite the difficulties we have in this field, it is my belief that it is a wonderful profession and a great way to serve the public.

Dr Shaun Brookhouse, April 17, 1998
Manchester, England

Appendix 1

The Foster Report

Chapter 9

Scientology and the Law

237. Our legal system today is rightly not concerned to restrict thoughts, beliefs, opinions or (with a few exceptions) the honest expression of any of these. Where the system does intervene is to restrain conduct for which there is evidence that it harms others. It is against this background that one has to consider the position of Scientology under our law.

(a) Therapy and claims to cure

238. That the practices of Scientology constitute a therapy, which claims to cure people of their real or imagined ills, must surely be beyond dispute. Many of the claims have already been quoted in earlier chapters of this Report, yet from time to time the Scientology leadership flatly denies that Scientology or Dianetics is a therapy. I have some difficulty in understanding how such denials can be put forward in the face of claims, from the same source, that "Dianetics is the most advanced and the most clearly presented method of psychotherapy and self-improvement ever discovered" (74) that "tiredness, unwanted sensations, bizarre pains and aches, bad hearing or sight...routinely respond to Dianetic processing" (75) or that "Scientology...has been remarkably effective in handling conditions and various mental states... Some 82 per cent of the clinical cases in the records of Scientology organisations show remarkable improvement in mental and physical condition. The records are meticulously kept and comprise the only validation programme of any therapy in Great Britain". (76)

239. Put bluntly, what is often said against the Scientology leadership is that they are quacks, dishonestly exploiting for their own financial gain the hopes of betterment or cure which they hold out to the anxious, the lonely, the inadequate, the credulous and the deluded, but in which they do not themselves believe.

240. For the reasons given in Chapter 2 of this Report, I have not come to any conclusion on the substance of these charges. Such charges are in any case notoriously difficult to prove, since they require proof of a state of mind which the person accused of them has every motive, if he is guilty, to dissimulate. It is only on rare occasions that, in an unguarded moment or perhaps within the inner circle of his co-conspirators, a confidence trickster will himself admit that he does not believe what he tells his victims. It is enough for me to say that there are in this Report quotations from the Scientology leadership's internal policy documents which display an attitude wholly different from that expressed to the public in general, and especially to potential

recruits. I have myself refrained from drawing any conclusions and have restricted myself to the publication of the relevant evidence.

(74) Cover of Dianetics: MSMH (1956).(75) HCO Bulletin of 24th April 1969, quoted in "The Auditor" No. 48.

(76) "Have you lived before this life?", p.16.

241. Even if the Scientology leadership were quacks, I doubt whether they would be committing any offence under English law as it stands. Telling lies is, by itself, no crime, unless the prosecution can prove a dishonest intention to obtain money, goods, services or sexual intercourse.

Nor is it any offence to claim to alleviate or even remove most ills (177): were it otherwise, no drug manufacturer who advertises his patent remedies could stay in business for long. (178)

242. The law does not forbid, in general terms, the practice of medicine or surgery by unqualified persons. The medical legislation makes provision, however, for the registration of persons who possess certain medical qualifications, and a person who practises medicine or surgery without being so registered is under serious disabilities as compared with a registered practitioner. Thus, he is forbidden to use any title or description implying that he is a registered practitioner or is recognised by law as a physician or surgeon; he is not entitled to recover in a court of law his charges for medical or surgical attendance or advice; he may not hold certain appointments which are closed except to registered practitioners; and he is not entitled to possess or supply dangerous drugs, and cannot give valid statutory certificates. On the other hand a person can only become a registered practitioner by acquiring a recognised qualification; and once registered he is subject to a strict disciplinary system in regard to his professional conduct. (179)

243. What the law does provide is a civil remedy, in that a patient who is treated with less than the due skill and care he is entitled to expect can recover damages for any harm which he suffers as a result. In the case of Scientology, the exemption clause in their standard contract which I have quoted in paragraph 135 above is designed to bar such an action, though I have some evidence that disenchanted pre-clears who have been forceful enough to sue the organisation for a return of their fees "on a consideration which has wholly failed" have had their cases successfully settled without the courts being troubled with a trial.

244. The Scientology leadership themselves told me that "the normal procedure when a person asks for a refund is for him or her to be given a refund with as little delay as possible. Only if the claim appears to be wholly unjustified on any grounds, do we contest the claim. There have been only two of these…"

245. The policy on which Parliament appears to have acted in the past is to control only those dangers which are not immediately obvious. Thus there is no law which prevents me from hiring a wholly unqualified surgeon to amputate my leg, but there is a law which prevents me from buying certain drugs at my chemist's without a prescription from a qualified medical man. This seems to me to make good sense: having my leg cut off is obviously a dangerous business, but the tiny pink pill which could kill me just as easily may look quite harmless in the palm of my hand.

(177) There are statutory exceptions for cancer and venereal disease in the Cancer Act 1939 and the Venereal Disease Act 1917.(178) There are only eight named diseases for the cure of which drugs must not be publicly advertised: Pharmacy and Medicine Act 1941.(179) Halsbury's Laws of England, 3rd ed., vol. 26, para. 2.

246. The question, therefore, which I have to consider is whether there is a case for legislation in the United Kingdom to control the practice of psychological medicine. I have come to the clear conclusion that there is, and that the case is a strong one. In what follows, I set out the arguments as they appear to me.

247. Let me begin by defining some terms. The human being likes to divide himself into a "body" and a "mind". Whether the division corresponds to any independent reality is a question which I prefer to leave to philosophers: it is enough to say here that not everyone agrees on where or how the dividing line should be drawn, but that medical men of most disciplines nowadays agree that, even if the two do have separate existences they interact strongly with each other. Some go so far as to say that all diseases are "psychosomatic", involving both the body and the mind. The mind in turn can usefully be described as behaving at times rationally and at other times irrationally: we are apt to describe the former as the functioning of our "intellect", and the latter as that of our "emotions". Those who study and treat our minds call themselves by a variety of names depending largely on the schools of theory and practice which they follow. For simplicity, and to avoid becoming entangled in technical detail, I shall label those who study the workings of (largely) our intellect "psychologists" and those who seek to alleviate or cure such of our illnesses as are thought to be of (mainly emotional) mental origin "psychiatrists". Psychologists often specialise in different branches of their subject, so that we have, for example, educational psychologists (who study how we learn), and industrial psychologists (who study us at work). Psychiatrists, broadly speaking, practise two distinct kinds of therapy: "physical" medicine, which seeks to affect our minds through our bodies by material interventions such as electric shocks or drugs; and "psychological" medicine, which seeks to affect our minds directly and without any material intervention (180). For this last technique I propose to use the expression "psychotherapy", regardless of the particular school or discipline – such as "psychoanalysis" or "analytical psychology" – which the therapist happens to follow. It will be immediately obvious that, in this terminology, Scientologists practise both psychology (in that they measure intelligence quotients and claim to improve them) and psychotherapy ("auditing" in particular and "processing" in general). This is indeed common

ground. Mr Hubbard himself describes Scientology and Dianetics as "that branch of psychology which treats human ability" (181) and as "the first thoroughly validated psychotherapy". (182)

248. Psychotherapy is a relatively new technique. Despite the often asserted proposition that it has been practised for centuries in the Roman Catholic confessional, its origin as a treatment for the relief or cure of illness is to be found with Professor Sigmund Freud.

It was he who first put forward, as a limited hypothesis subject to later disproof by the application of scientific method, the theory that our emotions go through certain stages of development in childhood, that their dynamics are predictable in broad terms, and that it is possible for a skilled therapist to intervene in those dynamics by a complex pattern of verbal communication with his patient on an emotional level. Whatever reservations may still be held on Freud's thesis, the techniques have been developed for the best part of three-quarters of a century and are practised today by tens of thousands of psychotherapists throughout the world.

(180) I am aware that all these expressions are frequently used in other senses: I am concerned here only to give convenient labels for the purpose of the discussion which follows. (181) Fundamentals of Thought, p.11. (182) Scientology 8-8008, p 90.

249. Enough is now known about the techniques of psychotherapy to establish the following propositions, with which 1 think few practising psychiatrists would disagree:–

(a) given the right conjunction of therapist and patient, psychotherapy can do much to relieve the latter's suffering;

(b) on the other hand, there are certain conditions (often recognisable only to an expert in the field) which respond very little, or not at all, to psychotherapy, whoever performs it;

(c) the techniques of psychotherapy are exceedingly complex and require great skill and long experience for their successful application;

(d) the possibilities of harm to the patient from the abuse, or the unskilled use, of these techniques are at least as great as the possibilities of good in the right hands.

250. One special aspect of psychotherapy requires mention here, and that is the so-called "transference" effect. From his earliest days, Freud observed that his patients were apt to transfer to him many of the emotions which had, for one reason or another, remained unresolved in their childhood, so that during the course of treatment he became the object of their most deeply seated feelings of love and hatred, of greed and generosity, of envy and gratitude, and often of sexuality. Such a situation

imposes a considerable strain on the therapist, and places a great weight of responsibility upon him. More than ever today, psychotherapists regard the ultimate dissolution of the transference at the end of the treatment as the most difficult, and yet the crucial, part of their task.

251. This very brief sketch of certain features of what has now become the principal technique in the armoury of modern psychological medicine (in my sense of the term) is necessarily over-simplified and incomplete, but I have said enough to explain why I have reached the conclusion that the intervention of Parliament has become necessary.

Here is a classic case of something which appears to the uninitiated as a wholly harmless procedure: all that you would see in a psychotherapist's consulting room is two people – or sometimes a group of people – talking to each other. Yet the danger in anything other than the most skilled hands is great and, what is worse, the possibilities of abuse by the unscrupulous are immense. The trained and selfless practitioner is concerned only to convert the deep emotional dependence on him which his patient develops during the treatment into an ability on the patient's part to wean himself from the therapist, and to achieve the maturity, and the independent ability to make relationships by choice, which are the aim of most of us. But it is fatally easy for the unscrupulous therapist, who knows enough to create the dependence in the first place, to exploit it for years on end to his own advantage in the form of a steady income, to say nothing of the opportunities for sexual gratification. While the latter would rapidly spell the end of a medically qualified therapist's practice at the hands of the General Medical Council, that body has no jurisdiction over therapists who do not happen to be doctors.

252. Further, it will not have escaped attention that those who feel they need psychotherapy tend to be the very people who are most easily exploited: the weak, the insecure, the nervous, the lonely, the inadequate, and the depressed, whose desperation is often such that they are willing to do and pay anything for some improvement of their condition.

253. In all this, there are analogies with a number of skilled activities which have been practised for much longer. Lawyers, doctors, architects and nurses, for example, all put at their client's service, for reward, intricate skills of which the clients are ignorant and which they must largely take on trust. All of them are conscious of the dependence which their professional relationships tend to create, and of the harm which they could do if they failed to use all their skill, or exploited the dependence in a selfish fashion.

254. The traditional method which we have used in this country – and which has been used in many others – to protect the weak from the exploitation which such a dependence makes possible, while ensuring that those skilled in their speciality can give of their best, is to create a controlled "profession". This involves the setting up of a body

(generally called a "Council") having authority over those who practise the particular skill concerned, leaving it to the Council to work out minimum standards of expertise for admission to the profession, a code of ethics and the exercise of disciplinary powers to enforce it, while the law places restraints of one sort or another upon the practice for reward of the particular skills concerned to those recognised as qualified by the Council.

255. Such a system has worked excellently in the past, and by and large the public has been well served by it. It ensures that those who are allowed to practise the skills without legal restriction have been properly trained in them, are fully aware of the moral responsibilities involved in their exercise, and that their continued livelihood depends on their continued discharge of those responsibilities.

According to a recent Report of the Monopolies Commission (183), there are at the present time in this country no fewer than 19 separate professions regulated in this fashion, whose members add up to over 850,000 people.

256. The price which we pay for such a system is some limitation on the opportunity to consult unqualified practitioners, and while in theory this restricts our freedom I doubt whether there are many who would, with all the relevant information at their disposal, take serious objection to this. Another disadvantage has sometimes been the resistance to innovation displayed by institutional bodies, but this is in my view heavily outweighed by the reduction of the risks involved in releasing revolutionary techniques on an unsuspecting public before their potential dangers have been fully investigated.

(83) Report on Professional Services (Cmnd. 4463), Part II, Table IV.

257. It may be instructive at this point to return for a moment to Sigmund Freud. In his lifetime, there was much debate in Austria on the question of whether the practice of psychoanalysis should be limited to qualified medical practitioners. Freud took the view that it should not – not because he thought that anyone and everyone could safely practise it, but because he did not think that a qualification in physical medicine was enough. In effect, he regarded psychoanalysis as a profession *sui generis*. In a book (184) devoted to the question, he said this:–

"No one should practise analysis who has not acquired the right to do so by a particular training. Whether such a person is a doctor or not seems to me immaterial" (185).

"The conditions will have to be laid down under which the practice of analysis shall be permitted to all those who seek to make use of it, an authority will have to be set up from whom one can learn what analysis is and what sort of preparation is needed for it, and the possibilities for instruction in analysis will have to be encouraged" (186).

"The analyst should possess personal qualities that make him trustworthy, and should have acquired the knowledge and understanding as well as the experience which alone can make it possible for him to fulfil his task" (187).

"The important question is not whether an analyst possesses a medical diploma but whether he has had the special training necessary for the practice of analysis" (188).

258. These, then, are the grounds on which I have become convinced that it is high time that the practice of psychotherapy for reward should be restricted to members of a profession properly qualified in its techniques, and trained – as all organised professions are trained – to use the patient's dependence which flows from the inherent inequality of the relationship only for the good of the patient himself, and never for the exploitation of his weakness to the therapist's profit. Such legislation already exists in a number of states in Europe, the Commonwealth and the United States.

259. That it is the phenomenon of Scientology which has pointed out this need in the existing law is a matter on which, if it is the leadership's sincere desire to help humanity, they will have cause to congratulate themselves.

Without coming to any conclusion on whether they in fact exploit their followers for their own profit, or whether it is desirable for auditors who may have had only a few weeks' training since they came to Scientology with problems of their own, to be encouraged to practise psychotherapeutic techniques on those who, *ex hypothesi*, are sitting targets for exploitation, the mere fact that such a situation could easily be abused at the present time with impunity demonstrates the urgent need for reform.

(184) The Question of Lay Analysis
(185) p.233
(186) p.238
(187) p.244
(188) pp.251 – 2

260. The details of the legislation which I recommend will need to be worked out by all parties concerned: Parliament, the relevant Departments, and the psychotherapists themselves. No doubt problems will arise, but I know of none which cannot be solved. In my view, the following are among the matters which will have to be borne in mind:–

(a) Clearly, it is only the practice of psychotherapy for fee or reward in cash or kind, paid by or on behalf of the patient, which needs to be controlled, since in a very wide sense we all practise some kind of psychotherapy on each other in our personal relationships, and many voluntary organisations try to help people with "counselling".

(b) I see no reason why (as in the recommendations of the Anderson Board (189)), doctors, dentists, ministers of religion, social workers and marriage guidance counsellors should be exempted from the provisions of this legislation: if any of these wish to charge their patients or clients for practising psychotherapy on them, there is no reason why they should not first satisfy the Council that they have undergone the necessary training and obtained the necessary qualifications.

(c) the new profession's governing body will need to have power to make transitional arrangements for the admission of persons with limited or even no formal qualifications if they can qualify on the grounds of past experience in the field and are willing to satisfy any necessary tests and submit themselves to the appropriate ethical code.

(d) when psychotherapy was first developed, its concepts were revolutionary and Freud had to contend with much opposition before his theories found general acceptance. The subject is young and still developing rapidly. Clearly, the new profession's rules will need to be more flexible than those adopted at the present time by, say, lawyers and accountants.

Had the medical profession been able to exclude osteopaths from practice in the past, much suffering might have gone unrelieved. It is therefore important to ensure that progress is not inhibited by the kind of conservatism which has, on occasions, tended to afflict some of the older professional bodies, particularly in the medical field. The best method of avoiding this pitfall is to provide for the appointment to the Council of a number of radically-minded laymen who will act as a leaven.

261. I see no reason at the present time why the practice of psychology (in my sense of the term) should be professionally restricted. The dangers inherent in an incompetent assessment of someone's intellectual capabilities or his fitness for a particular employment, albeit regrettable, do not appear to me to be of a comparable order with those resulting from an abuse, or an incompetent use, of a system of therapy which operates by a deliberate intervention in the patterns of people's irrational emotions.

(189) Anderson Report, p.171.

262. Finally, I should say that I disagree profoundly with the legislation adopted in both Western and South Australia, in turn based on part of that adopted in Victoria, whereby the teaching and practice of Scientology as such is banned. Such legislation appears to me to be discriminatory and contrary to all the best traditions of the Anglo-Saxon legal system. I cannot see any reason why Scientologists should not be allowed to practise psychotherapy if they satisfy the proposed professional body that they are qualified to do so, that their techniques are sound, that their practitioners receive adequate training and operate under a stringent ethical code, and that there is no hint of exploitation. If it is indeed, as they claim, "the first thoroughly validated psychotherapy", the profession will welcome them with open arms. And should its governing body decide, as has been done in many professions, that it is unethical to

advertise for patients or to make unqualified claims to cure, I have no doubt that the Scientology leadership, if its sincerity is genuine, will be happy to conform to these standards.

(b) The privileged position of religious bodies

263. One other matter of substance has arisen in the course of this Enquiry which, in my view, merits further consideration, and that is the variety of privileges which the laws of this country confer upon associations of mortals who combine for religious purposes. The source of these privileges – some of which are of substantial economic value – is to be found in the remoter parts of our history, and it appears to me to be debatable what correlative benefit our society today derives from their continued existence.

264. The privileges themselves are numerous, and occur sporadically and without much logic in a number of areas. The more important ones are these:

(1) "The advancement of religion" is one of the three main purposes which validates a "charity" in law – the others being the advancement of education and the relief of the poor.

Charitable bodies (i.e. trusts or corporations) enjoy the advantage of exemption from income tax and surtax (190), and Capital Gains Tax (191), to the extent to which their income or gains are applied for charitable purposes; gifts to them are not void for perpetuity or inalienability as they would be if made to anyone else; and charitable corporations may be exempted by the Board of Trade from adding the word "Limited" to their names (192). Nor do charities pay Selective Employment Tax (193), so that the domestic servants of a minister of religion, if they are employed by his Church, do not attract this levy while most other domestic servants do.

(2) By Statute (194) a place may be registered as a place of religious worship, and if it is it will be exempt from all rates (195) and contributions to roadworks (196) and sewers (197).

(190) Income and Corporation Taxes Act 1970, Section 360.
(191) ibid, and Finance Act 1965, Section 35(1).
(192) Companies Act 1948, Section 19(1).
(193) Selective Employment Payments Act 1966, Section 5.
(194) Places of Worship Registration Act 1855.
(195) Rating and Valuation Act 1925, Sections 2(3) and 3(2).
(196) Private Street Works Act 1892, Section 16.
(197) Public Health Act 1875, Section 151.

265. Whether or not it may be thought desirable to continue to confer these privileges on bona fide religions having a substantial following, there seems to me to be a clear need for precautions which will ensure that there can be no abuse. As matters stand, it is enough for any small group of people to come together and claim to believe in, and worship, a deity, and this is clearly not good enough in the light of the great economic value of the privileges concerned.

266. In these circumstances, I recommend that the time is ripe for a review of the law which accords these privileges to religious bodies, with the object of at least ensuring that they are restricted to religious movements having a substantial number of adherents, and engaging in genuine acts of worship.

267. Any such review of the law must obviously not detract in any way from the existing tolerance of religious belief, whether it be Christian, Jewish, infidel or heathen, but should confine itself to regulating more strictly the fiscal exemptions which religious bodies can enjoy.

(c) Miscellaneous

268. Two further matters deserve mention. First, I am struck by the ease with which "non-profit-making" companies or associations are able to escape the payment of taxes, even if they are not charities. An ordinary business pays tax on the whole of its income, after deducting only those expenses incurred "wholly and exclusively" for the purpose of the business, and the Inland Revenue authorities not unnaturally subject these expenses to close scrutiny.

But in the case of an organisation which renders paid services only to its members, the system is different: a principle of "mutuality" is applied, with the result that the full income from the members (in the form of fees) escapes taxation at that point, and so do donations from non-members. Moreover, if the organisation then distributes its surplus by way of donations to associated companies, or even to individuals, these payments are still not liable to tax because they are "voluntary" payments. If the services were sold to the general public who are not "members", such an organisation would have to pay taxes like everyone else, and only legitimate business expenses would be deductible; but considering the ease with which one can enrol "members", the distinction strikes me as artificial. This aspect of our tax system is in my opinion ripe for review.

269. Payments such as those shown in the Scientology Companies' accounts as being made to other Scientology organisations, or to Mr or Mrs Hubbard, who are not residents of the sterling area, of course require the consent of the Bank of England under the Exchange Control Act.

270. The other matter which deserves attention is the failure of a number of the Scientology companies to file accounts and annual returns within the time prescribed by the law, without apparently incurring any sanction at the hands of the Registrar of

Companies. These sanctions seem to me pointless if they are not enforced.

Appendix 2

Shaun Brookhouse, PhD, DCH, MSc, PGDHP, FCAP, FASH, FHRS
UKCP Registered Hypno-Psychotherapist
Richmael House, 25 Edge Lane, Chorlton cum Hardy, Manchester M21 9JH
Tel: 0161 881 1677 Fax: 0161 882 0376
e-mail: DrBrookhouse@hypnotherapy.demon.co.uk
http://www.hypnotherapy.demon.co.uk

UK Training College
St Charles Hospital
Exmoor St
North Kensington
London
W10 6DZ

16 September, 1997

Dear Mandy:

I am writing to ask your help. I am currently researching an MA in Educational Studies at Liverpool John Moores University. The title of my thesis is *Hypnotherapy Training In The UK: An Investigation Into The Development Of Clinical Hypnosis Training Post-1971*.

The help that I need from you is a copy of your prospectus, that I can use for research purposes. As you are probably aware, I too am a hypnotherapy trainer with a school of my own, but the purpose of my request has nothing to do with this role or the school. It is purely for academic purposes that I request your prospectus.

If you could see your way clear to sending it, I would be most grateful. Your organisation will, of course, get a mention in the final thesis.

Thanking you in anticipation.

Yours sincerely,

Dr Shaun Brookhouse
Hypno-Psychotherapist

Appendix 3

Shaun Brookhouse, PhD, DCH, MSc, PGDHP, FCAP, FASH, FHRS
UKCP Registered Hypno-Psychotherapist
Richmael House, 25 Edge Lane, Chorlton cum Hardy, Manchester M21 9JH
Tel: 0161 881 1677 Fax: 0161 882 0376
e-mail: DrBrookhouse@hypnotherapy.demon.co.uk
http://www.hypnotherapy.demon.co.uk

15 January, 1998

Dear Colleague:

In September, 1997, you kindly replied to my request for your training details in Clinical Hypnosis. Thank you very much for that.

I would like to impose on you once again, and for the last time, to ask if you could complete the enclosed questionnaire. It is only ten questions and has been laid out so it should not take more than a couple of minutes of your time.

I would appreciate it if you could return it to me in the enclosed stamped addressed envelope no later than 31 January, 1998.

Thanking you for all of your kind assistance.

Yours sincerely,

Dr Shaun Brookhouse
Hypno-Psychotherapist

Hypnotherapy Training

Questionnaire

1. Strongly Agree   2. Agree   3. No opinion   4. Disagree   5. Strongly Disagree

i.   The state of hypnosis training in the UK is fine and clear to the public.   _____

ii. Hypnotherapy practice should be limited to graduates of medicine, dentistry, and/or psychology.   _____

iii. Legislation is required to regulate the training and practice of hypnosis.   _____

iv. Legislation is likely in the next ten years to cover Q3.   _____

v.   Your training would recognise graduates from other colleges.   Yes/No

vi. Your training has some form of University validation.   Yes/No

vii. Voluntary self-regulation would be a useful system of quality control for hypnosis training.   _____

viii. NVQs are a positive development for hypnosis training.   _____

ix. Correspondence courses provide ample training for hypnotherapists.   _____

x.   Hypnotherapy is a profession in its own right.   _____

The name of your organisation: _____

Your position/title in that organisation: _____

Please return to Dr S. Brookhouse using the Stamped Address Envelope enclosed no later than 31 January, 1998. Thank you for your assistance!

Appendix 4

Complaints and Disciplinary Procedure for The Centre Training School of Hypnotherapy and Psychotherapy and The Centre Association of Psychotherapists

All complaints against members, other than those being dealt with by Courts of Law, will be dealt with by the advisory and governing council of the Association, or on the recommendations of any disciplinary body established under its authority (hereinafter called the Disciplinary Committee).

1. The Disciplinary Committee shall consist of a minimum of 3 persons, including the chairperson.

2. On receiving a complaint against any of its practitioner members, the CAP Secretary will immediately request from the complainant that full details of the complaint be submitted in writing (if this has not already been done). The complainant must be informed, also in writing, that although the Association's Disciplinary Committee is prepared to respect confidentiality and to note the nature of the complaint, it cannot proceed against any therapist unless the full nature of the complaint and the name of the complainant be made known to the therapist.

3. If the complainant wishes to proceed with the matter, the Secretary will then, in writing, within 14 days, inform the therapist concerned, giving full details and inviting the therapist to make any written comments that he/she wishes.

4. Having given reasonable them for a reply to be received – no more than 21 days – the Secretary will then place all the details before the Disciplinary Committee.

5. Should the Disciplinary Committee consider that a *prima facie* case has been established, it will make arrangements for a hearing of the case within a period of four weeks.

6. The Secretary will inform both the complainant and the therapist of this hearing, inviting them to attend in person if they wish to do so and bring along any witnesses, and if they so wish, be professionally represented.

7. If either the member or complainant called to such a meeting fails to appear, the Disciplinary Committee may make a decision *in absentia*.

IF EITHER COMPLAINANT OR THERAPIST SEEKS TO TALK TO ANY MEMBER OF THE ASSOCIATION'S COUNCIL OR DISCIPLINARY COMMITTEE ABOUT THE COMPLAINT PRIOR TO THE MEETING, THEY SHOULD REFUSE TO RESPOND.

8. The Disciplinary Committee shall convey its findings, within 21 days, to the organisation's council, who will accept these findings, and in the light of such, decide upon any further action to be taken. Possible outcomes are as follows:

complaint is not upheld – no action

complaint is upheld – written reprimand is given

complaint is upheld – membership is terminated

ONCE THE DISCIPLINARY COMMITTEE HAS DECIDED TO UPHOLD A COMPLAINT, BUT BEFORE DECIDING UPON THE COURSE OF ACTION, IT WILL HAVE INFORMATION LAID BEFORE IT BY THE ASSOCIATION'S COUNCIL OF OUTCOMES OF ANY PREVIOUS DISCIPLINARY HEARINGS IN RESPECT OF THE DEFENDANT.

9. All decisions will be conveyed to both complainant and defendant in writing within 21 days.

IN THE EVENT OF EITHER PARTY NOT BEING SATISFIED WITH THE PROCEDURES RESULTING IN THE DECISION ARRIVED AT, THE SECRETARY WILL INFORM BOTH COMPLAINANT AND THE THERAPIST THAT ALL WRITTEN MATERIAL PERTAINING TO THE COMPLAINT IS BEING PASSED TO THE NATIONAL BODY OF WHICH CTSHP AND CAP ARE MEMBERS.

Appendix 5

Code of Ethics and Practice
National Council for Hypnotherapy

1. Provide a service to clients solely in those areas in which they have demonstrated competence and for which they carry full professional indemnity insurance that is acceptable to NCH.

2. Be aware of their own limitations and wherever appropriate, be prepared to refer a client to another suitable practitioner (whether or not such practitioner be a member of the NCH) who might be expected to offer suitable treatment.

3. Ensure, as far as reasonably possible, that wherever an aspect of the client's condition is either known or suspected, the treatment of which is beyond their training and expertise, the client be advised to seek medical or other appropriate advice.

4. Maintain strict confidentiality within the relationship consistent with the good care of the client and the laws of the land and ensure client notes and records be kept secure and confidential. (Members should be aware that in the event of a complaint against them, subject to the complainant providing written consent for their notes and records to be made available, members may be required to provide this data to the Committee on demand.)

5. Obtain written permission from the client (or if appropriate, the client's parent/s or guardian/s) before either recording client sessions or discussing undisguised cases with any person whatsoever. ("Recording" in this context means any method other than the usual taking of written case notes. "Undisguised" in this context means cases in which material has not been sufficiently altered in order to offer reasonable anonymity to all relevant parties).

6. Take all reasonable steps to ensure the safety of both the client and any person who may be accompanying the client.

7. Refrain from using their position of trust and confidence to:

   a) Exploit the client emotionally, sexually, financially or in any other way whatsoever. Should either a sexual or financial relationship (ie other than the payment of session fees, or the purchase of books, tapes, or other relevant products) develop between either therapist and client or members of their respective immediate families, the therapist must immediately cease to accept fees, terminate treatment consistent with Clause 10 and transfer the client to another suitable therapist at the very earliest opportunity. (Clarification on this can be offered by the Committee via the Secretary.)

b) Touch the client in any way that may be open to misinterpretation. (It must be realised, that notwithstanding the above, any evidence produced against a member of the NCH, showing they have abused their position of trust to exploit a client, will result in the maximum penalty the NCH can impose.)

8. Not permit considerations of religion, nationality, gender, age, disability, politics, or social standing to influence, adversely, client treatment.

9. Ensure that their workplace and all facilities offered to both clients and their companions will be in every respect suitable and appropriate for the service provided.

10. Terminate treatment at the earliest moment consistent with the good care of the client.

11. Disclose full details of all relevant training, experience and qualifications to clients, upon request.

12. Make no claim that they hold specific qualifications unless such claim can be fully substantiated. In the absence of qualifications permitting practice in the UK as a medical doctor, should the title of "Doctor", to which a member has legal entitlement, be utilised within a therapeutic context, clients must be suitably advised, at the very earliest opportunity, that the member is not medically qualified. Additionally, all relevant correspondence (however generated), advertising, promotional material and business stationery, the publication of which has involved the member themselves and which incorporates the non-medical, yet legally held title of Doctor, must bear a prominent statement (i.e. of minimum type size 10 pt Courier or its equivalent and with regard to business cards, a minimum type size of 6 pt Courier or its equivalent) that the member is not medically qualified.

13. Use no claim or title connected with the National Council other than that they are members of the NCH with appropriate designated letters. (Officers of the Council shall be further entitled to state the position held by them.)

14. Explain fully to clients in advance of any treatment, the fee levels, practice terms of payment and any charges which might be imposed for non-attendance or cancelled appointments.

15. Present all services and products in an unambiguous manner and ensure that the client retains complete control over the decision to purchase such services or products.

16. Conduct themselves at all times in accord with their professional status, and in such a way as neither undermines public confidence in the process or profession of hypnotherapy nor brings the reputation of the NCH into disrepute.

17. a) Confirm that they have not been in any disciplinary action before any professional body, neither has membership of any such association been terminated nor application rejected, and undertake to inform the NCH should such an event either take place or have taken place.

    b) Confirm that they have not been convicted of any offence likely to bring their professional name or the reputation of the NCH into disrepute, and undertake to inform the Secretary should such an event either take place or have taken place.

18. Inform the Secretary of any alteration in circumstances which would affect either their position or ability as practising members.

19. Notify the Secretary in writing of any changes in contact address and/or telephone number at the earliest convenience.

20. Ensure that all advertising shall comply with the British Code of Advertising Practice, accord with the British Advertising Standards Authority and make available all such literature to the Committee on demand.

21. Accept that this is not a static document and that it may be altered, from time to time, by the Committee of Management of the NCH, in accordance with the need to ensure ongoing professionalism within the field of hypnotherapy.

Bibliography

Ahier, John, Cosin, Ben, & Hales, Margaret (Eds.):
Diversity And Change Education, Policy, And Selection, Routledge, London, 1996.

Alman, Brian & Lambrou, Peter:
Self Hypnosis: The Complete Manual For Health And Self Change, Brunner Mazel, New York, 1992.

American Institute Of Hypnotherapy:
Institute Catalogue, ABH Press, Irvine, California, 1997.

American Board of Hypnotherapy
Membership Brochure, ABH Press, Irvine California, 1998

American Council of Hypnotist Examiners,
The Education and Training of Hypnotherapists, Westwood Publishing, Glendale California 1998

American Pacific University
University Catalogue, AND Publishers, Honolulu Hawaii 1999

Bell, Leslie:
On Becoming A Research Student, John Moores University, Liverpool, 1997.

Berg, Morris & O'Sullivan, Michael:
Hypnotherapy Resources And Career Guide, Symbolon Press, London, 1997.

Boye-Thompson, Titus:
The UK Alternative And Complementary Medicines Handbook, Almedic Publishing, London, 1996.

British Medical Association:
Complementary Medicine New Approaches To Good Practice, Oxford University Press, Oxford, 1993.

British Medical Association:
"Psychological Medicine Group Sub-Committee", *British Medical Journal Supplement*, London, 1955.

British Society Of Medical And Dental Hypnosis:
Accreditation Guidelines, Rotherham, 1997.

Brookhouse, Shaun
Opening Address to the Formation Meeting of UKCHO, London 1998

Brookhouse, Shaun:
"Legislative Opinions", *The Journal Of The National Council Of Psychotherapists And Hypnotherapy Register*, Summer Edition, 1995.

Brookhouse, Shaun:
"The State Of The Union Is Now Important For Hypnotherapists", *The European Journal Of Clinical Hypnosis*, Volume 1, Number 3, 1994.

Brookhouse, Shaun:
Brookhouse Hypnotherapy Brochure, Manchester, 1997.

Broom, William:
"Secretary's Report To The National Council For Hypnotherapy", *The Journal Of The National Council Of Psychotherapists And The National Council For Hypnotherapy*, Spring Edition, 1997.

Care Sector Consortium:
National Occupational Standards For Complementary Medicine Final Report, London, 1997.

Centre For Psychotherapeutic Studies:
Course Prospectus, Sheffield University, Sheffield, 1997.

Centre Training School Of Hypnotherapy And Psychotherapy:
Course Prospectus, Preston, 1997.

Cousins, Anne:
"President's Report To The National Council Of Psychotherapists And Hypnotherapy Register", *Journal Of The National Council Of Psychotherapists And Hypnotherapy Register*, Winter Edition, 1995.

Damon, Dwight:
"The Editor Speaks", *The Journal Of Hypnotism*, Volume 12, Number 1, 1997.

Dey, I:
Qualitative Data Analysis, Routledge, London, 1993.

Erickson, Milton H.:
The Collected Papers Of Milton H. Erickson, MD, Volume 1, Irvington Publishers, New York, 1980.

Ewing, Dabney:
"Pain Management Seminar Bury Postgraduate Medical Centre", Bury, 1996.

Foster, Sir John:
   Enquiry Into The Practice And Effects Of Scientology, Her Majesty's Stationery Office, London, 1971.

Gittens, Cynthia:
   "The Need For Choice", *Newscheck*, Volume 6, Number 4, 1996.

Gomm, Roger & Woods, Peter:
   Educational Research In Action, Volume 2, Paul Chapman Publishing, London, 1993.

Heap, Michael & Dryden, Windy:
   Hypnotherapy: A Handbook, Open University Press, Buckingham, 1991.

Hypnotherapy Research Society:
   Prospectus, Stone in Oxney, 1997.

Hypnotism Act:
   An Act To Regulate The Demonstration Of Hypnotic Phenomena For Purposes Of Public Entertainment, HMSO, London, 1952.

Institute For Complementary Medicine:
   An Invitation to Join, London, 1991.

International Association Of Hypno-Analysts:
   Prospectus, Bournemouth, 1997.

Kinnoull, The Earl of:
   "Hypnotism Bill House Of Lords Second Reading", *House Of Lords Official Report 1541*, London, HMSO, London, 1979.

Mills, Simon & Peacock, Wendell:
   University Of Exeter Professional Organisation Of Complementary And Alternative Medicine In The UK 1997: A report to the Department of Health, Centre for Complementary Health Studies, Exeter, 1997.

Morgan, Dylan:
   "Survey of Parliamentary Opinion On Hypnotherapy", *The Journal Of The National Council Of Psychotherapists And Hypnotherapy Register*, Autumn Edition, 1995.

Morgan, Dylan:
   "Editorial", *The Journal Of The National Council Of Psychotherapists And Hypnotherapy Register*, Winter Edition, 1995.

National Guild of Hypnotists
"1998 Year End Report", *Journal of Hypnotism*, December 1998

National Guild of Hypnotists
   The 1995 Convention Manual, National Guild of Hypnotists, Merrimack New Hampshire, 1995

National School Of Hypnosis And Psychotherapy:
   Guide to Training and Registration, London, 1997.

New Hampshire State Legislature:
   HB 1508 FN An Act to Regulate Hypnotherapy, Merrimack, 1997.

Psychotherapy And Hypnosis Training Association:
   Prospectus, London, 1997.

Tart, C., (Ed.):
   Altered States Of Consciousness: A Book of Readings, John Wiley and Sons, New York, 1969.

UKCHO
   "*Constitution*" UKCHO London 1998

UKCP:
   "Guidelines To Registration", *National Register Of Psychotherapists*, Routledge, London, 1997.

UK Training College for Complementary Health Care Study:
   Course Prospectus, London, 1997.

Van Durzen, Emmy:
   Debate on the European Certificate of Psychotherapy, UKCP AGM And Conference, Wakefield, 1998.

Vetere, Arlene, Newell, Adrian, Watts, Mary, & Kosuiner, Adele:
   "The Registration of Psychologists Specialising in Psychotherapy", *The Psychologist*, Volume 10, Number 6, 1997.

Washington School of Clinical and Advanced Hypnosis:
   Prospectus, Manchester, 1997.

Waxman, David (Ed.):
   Hartland's Medical And Dental Hypnosis 3rd Edition, Bailliere Tindall, London, 1989.

Woodbury Counselling:
   "Dr. Chris Harvey Testimonial", *Course Prospectus*, Tenterden, 1997.

Yapko, Michael:
Trancework: An Introduction To The Practice Of Clinical Hypnosis 2nd Edition, Brunner Mazel, New York, 1990.

Yin, R.K.:
Case Study Research: Design And Method, Sage Publications, London, 1989.

Crown House Publishing Limited
Crown Buildings,
Bancyfelin,
Carmarthen, SA33 5ND
Wales.
Telephone: +44 (0) 1267 211345 Facsimile: +44 (0) 1267 211882
e-mail: books@crownhouse.co.uk Website: www.crownhouse.co.uk

We trust you enjoyed this title from our range of bestselling books for professional and general readership. All our authors are professionals of many years' experience, and all are highly respected in their own field. We choose our books with care for their content and character, and for the value of their contribution of both new and updated material to their particular field. Here is a list of all our other publications.

Communication Excellence: Using NLP To Supercharge Your Business Skills by Ian R. McLaren	Paperback	£12.99
Change Management Excellence: Putting NLP To Work In The 21st Century by Martin Roberts PhD	Hardback	£25.00
Dreaming Realities: A Spiritual System To Create Inner Alignment Through Dreams by John Overdurf CAC & Julie Silverthorn MS	Paperback	£9.99
Ericksonian Approaches: A Comprehensive Manual by Rubin Battino MS & Thomas L. South PhD	Hardback	£35.00
Ericksonian Approaches: Exercises & Demonstrations by Rubin Battino MS & Thomas L. South PhD	Audiotape	£9.99
Excel with the Multiple Intelligences Poster Set by Matthew Pearce	Nine posters	£19.99
Figuring Out People: Design Engineering With Meta-Programs by Bob G. Bodenhamer DMin & L. Michael Hall PhD	Paperback	£18.99
First Steps: To A Physical Basis Of Concentration by Roy Anderson	Paperback	£12.99
Gold Counselling, Second Edition: *A Structured Psychotherapeutic Approach To The Mapping And Re-Aligning Of Belief Systems* by Georges Philips & Lyn Buncher with Brian Stevenson	Paperback	£16.99
Grieve No More, Beloved: *The Book Of Delight* by Ormond McGill	Paperback	£12.99
Guided Imagery And Other Approaches To Healing by Rubin Battino MS	Hardback	£25.00
Hypnosis: A Comprehensive Guide by Tad James MS, PhD with Lorraine Flores & Jack Schober	Hardback	£20.00
Influencing With Integrity: Management Skills For Communication & Negotiation by Genie Z Laborde PhD	Paperback	£14.99
Instant Relaxation: How To Reduce Stress At Work, At Home And In Your Daily Life by Debra Lederer & L. Michael Hall PhD	Paperback	£8.99
The Magic Of Mind Power: Awareness Techniques For The Creative Mind by Duncan McColl	Paperback	£8.99
Me, Myself, My Team: How To Become An Effective Team Player Using NLP by Angus McLeod PhD	Paperback	£12.99
A Multiple Intelligences Road To An ELT Classroom by Michael Berman	Paperback	£19.99

Multiple Intelligences Poster Set by Jenny Maddern	Nine posters	£19.99
The New Encyclopedia Of Stage Hypnotism by Ormond McGill	Hardback	£45.00
Now It's YOUR Turn For Success! Training And Motivational Techniques For Direct Sales And Multi-Level Marketing by Richard Houghton & Janet Kelly	Paperback	£9.99
Peace Of Mind Is A Piece Of Cake by Michael Mallows & Joseph Sinclair	Paperback	£8.99
The Power Of Metaphor: Story Telling & Guided Journeys For Teachers, Trainers And Therapists by Michael Berman & David Brown	Paperback	£12.99
The POWER Process: An NLP Approach To Writing by Sid Jacobson & Dixie Elise Hickman	Paperback	£12.99
Precision Therapy: A Professional Manual Of Fast And Effective Hypnoanalysis Techniques by Duncan McColl	Paperback	£16.99
Principled Headship: A Teacher's Guide To The Galaxy by Terry Mahony	Paperback	£9.99
Rapid Cognitive Therapy, Volume 1: The Professional Therapists' Guide To Rapid Change Work by Georges Philips & Terence Watts	Hardback	£20.00
Scripts & Strategies In Hypnotherapy: The Complete Works by Roger P. Allen	Hardback	£25.00
Seeing The Unseen: A Past Life Revealed Through Hypnotic Regression by Ormond McGill	Paperback	£12.99
Smoke-Free And No Buts! by Dr Geoff Ibbotson & Dr Ann Williamson	Paperback	£5.99
Solution States: A Course In Solving Problems In Business With The Power Of NLP by Sid Jacobson	Paperback	£12.99
The Sourcebook Of Magic Second Edition: A Comprehensive Guide To The Technology of NLP by L. Michael Hall PhD & Barbara Belnap	Paperback	£18.99
The Spirit Of NLP Revised Edition: The Process, Meaning And Criteria For Mastering NLP by L. Michael Hall PhD	Paperback	£18.99
Sporting Excellence: Optimising Sports Performance Using NLP by Ted Garratt	Paperback	£12.99
Still – In The Storm: How To Manage Your Stress And Achieve Balance In Life by Dr Ann Williamson	Paperback	£5.99
The User's Manual For The Brain Volume I: The Complete Manual For Neuro-Linguistic Programming Practitioner Certification by Bob G. Bodenhamer DMin & L. Michael Hall PhD	Hardback	£29.50
Warriors, Settlers & Nomads: Discovering Who We Are & What We Can Be by Terence Watts	Paperback	£12.99

USA & Canada orders to:
Crown House Publishing
P.O. Box 2223, Williston, VT 05495-2223, USA
Tel: 877-925-1213, Fax: 802-864-7626
E-mail: info@CHPUS.com
www.CHPUS.com

UK & Rest of World orders to:
The Anglo American Book Company Ltd.
Crown Buildings, Bancyfelin, Carmarthen, Wales SA33 5ND
Tel: +44 (0)1267 211880/211886, Fax: +44 (0)1267 211882
E-mail: books@anglo-american.co.uk
www.anglo-american.co.uk

Australasia orders to:
Footprint Books Pty Ltd.
Unit 4/92A Mona Vale Road, Mona Vale NSW 2103, Australia
Tel: +61 (0) 2 9997 3973, Fax: +61 (0) 2 9997 3185
E-mail: info@footprint.com.au
www.footprint.com.au

Singapore orders to:
Publishers Marketing Services Pte Ltd.
10-C Jalan Ampas #07-01
Ho Seng Lee Flatted Warehouse, Singapore 329513
Tel: +65 6256 5166, Fax: +65 6253 0008
E-mail: info@pms.com.sg
www.pms.com.sg

Malaysia orders to:
Publishers Marketing Services Pte Ltd
Unit 509, Block E, Phileo Damansara 1, Jalan 16/11
46350 Petaling Jaya, Selangor, Malaysia
Tel : 03 7955 3588, Fax : 03 7955 3017
E-mail: pmsmal@po.jaring.my
www.pms.com.sg

South Africa orders to:
Everybody's Books
PO Box 201321, Durban North, 4016, RSA
Tel: +27 (0) 31 569 2229, Fax: +27 (0) 31 569 2234
E-mail: warren@ebbooks.co.za

Appointed EU Representative: Easy Access System Europe Oü, 16879218
Address:Mustamäe tee 50, 10621, Tallinn, Estonia
Contact Details: gpsr.requests@easproject.com,
+358 40 500 3575

www.ingramcontent.com/pod-product-compliance
Ingram Content Group UK Ltd.
Pitfield, Milton Keynes, MK11 3LW, UK
UKHW051524180426
11947UKWH00018B/1552